FOOD PROCESSING
(FRUITS OF INDIA)

Dr. Devendra Kumar Bhatt

SHREE PUBLISHERS & DISTRIBUTORS
NEW DELHI-110 002

Edition : 2013

Published by :
SHREE PUBLISHERS & DISTRIBUTORS
22/4735 Prakash Deep Building,
Ansari Road, Darya Ganj,
New Delhi–110 002

ISBN : 978–81–8329–511–6

Printed by :
Anvi Composers
Delhi

Preface

The present book on Fruit Processing is intended as a resource for students and researchers working in the field of Food Processing. Its broad scope makes it a useful volume which tries to fill the lacuna in the existing literature on food processing.

The structure and goal of the book combines the various principles involved in food processing as well as the details of fruits which are found in the Indian subcontinent.

The book has been divided into two sections. The first four chapters cover background information of food processing with special emphasis on fruits grown in India. The next fourteen chapters have discussed the processing of fourteen different fruits ranging from apple, guava, banana, oranges, amla, bel, jamun etc.

Man has exhibited much thought and foresight in cultivating a variety of grains, fruits, vegetables, nuts and oilseeds and in rearing birds and animals for use as food. All the mans food are perishable commodities, they begin to deteriorate shortly after harvest, gather or storage. A good knowledge its properties is required for its effective use. Food Science and Technology can help in the best utilization of the food with the changing lifestyle. India is one of the major producer of fruits and vegetables in the world and more than 40 percent of fruits and vegetables are wasted due to improper handling and post harvest conditions. There is need for the development of new processes and technologies for the proper post harvest handling and processing of fruits and vegetables so that most of the produce will be utilized and India will be able to increase its share in world processed food market where Indian share is less than 1.5 percent.

I hope that the book will be beneficial for the students and researchers working in this field.

—Dr. Devendra Kumar Bhatt

Contents

SECTION ONE
INTRODUCTION TO FOOD PROCESSING

Chapter 1

An Introduction to Food Processing

Food Industry is one of the biggest and fastest growing industries in the world. As the population is growing and land area is decreasing there is need for better food processing techniques and processes and more technical persons / food technologists. There is need for a lot of research so that there will be decrease in the post harvest losses and production of better quality products. This will increase the scope of food science and technology.

Food Science is a science that deals with the physical, chemical and biological properties of foods as they relate to stability, cost, quality, processing, safety, nutritive value, wholesomeness and convenience. It is an interdisciplinary subject involving primary bacteriology, chemistry, biology and Engineering. Food chemistry deals with composition, and properties of foods and chemical changes it undergoes during processing, handling and storage. Food chemistry is intimately related to chemistry, biochemistry, physiological chemistry, botany, zoology and molecular biology. Food chemists are concerned primarily with biological substances that are dead or dying (post harvest physiology of plant and post mortem physiology of muscles) and changes they undergo when exposed to a very wide range of environmental condition. The origin of food chemistry extends to antiquity, the most significant discoveries began in the late 1700's. The best accounts of developments during the period are of Filby and Browne. Food chemists are concerned primarily with biological substances that are dead or dying (post harvest physiology of plant and post mortem physiology of muscles) and changes they undergo when exposed to a very wide range of environmental condition.

The study of food science emphasises on the composition of foods and the changes that occur when they are subjected to food processing. Functional foods are foods that promote health beyond providing basic nutrition. Our nutritional status, health, physical and mental faculties depend on the food we eat and how we eat it.

Access to good quality food has been man's main endeavour from the earliest days of human existence. Safety of food is a basic requirement of food quality. "Food safety" implies absence or acceptable and safe levels of contaminants, adulterants, naturally occurring toxins or any other substance that may make food injurious to health on an acute or chronic basis. Food quality can be considered as a complex characteristic of food that determines its value or acceptability to consumers. Besides safety, quality attributes include: nutritional value; organoleptic properties such as appearance, colour, texture, taste; and functional properties. Food systems in developing countries are not always as well organised and developed as in the industrialised world. Moreover, problems of growing population, urbanisation, lack of resources to deal with pre- and post- harvest losses in food, and problems of environmental and food hygiene mean that food systems in developing countries continue to be stressed, adversely affecting quality and safety of food supplies.

Scheele, 1742-1786 (Swedish scientist) isolated and studied the properties of lactose, prepared music acid by oxidation of lactic acid (1780), devised a means of preserving vinegar, by means of heat, isolated citric acid from lemon juice (1784). Antoine Lavoiser (french chemist), 1743-1794, established fundamental principles of combustion, and organic analysis, attempted elemental analysis of alcohol (1784), presented first papers on organic acids of various fruits.

Liebig (1803-1873) showed in 1837 that acetaldehyde occurs as intermediate between alcohol and acetic acid during fermentation of vinegar. He classified foods into

- nitrogenous (vegetable, albumin, casein, and animal flesh and blood)
- non-nitrogenous (fats, carbohydrates, and alcoholic beverages).

There are four components to analytical approach to the chemistry of food formulation processing and storage stability

1. Determining those properties that are important characteristics of safe high quality foods.
2. Determining those chemical and biochemical reactions that have important influences on loss of quality and / or wholesomeness of foods.
3. Integrating the first two points so that one understands how the key chemical and biochemical reactions influences quality and safety.
4. Applying the understanding to various situations encountered during formulation, processing and storage of foods.

Reactions causes alterations in foods. Deterioration of food usually consists of a series of primary events followed by secondary events, which in turn becomes evident as altered quality attribute.

In developing countries agriculture is the mainstay of the economy. As such, it should be no surprise that agricultural industries and related activities can account for a considerable proportion of their output. Of the various types of activities that can be termed as agriculturally based, fruit and vegetable processing are among the most important.

Both established and planned fruit and vegetable processing projects aim at solving a very clearly identified development problem. This is that due to insufficient demand, weak infrastructure, poor transportation and perishable nature of the crops, the grower sustains substantial losses. During the post-harvest glut, the loss is considerable and often some of the produce has to be fed to animals or allowed to rot.

Even established fruit and vegetable canning factories or small/medium scale processing centres suffer huge loss due to erratic supplies. The grower may like to sell his produce in the open market directly to the consumer, or the produce may not be of high enough quality to process even though it might be good enough for the table. This means that processing capacities will be seriously underexploited.

The main objective of fruit and vegetable processing is to supply wholesome, safe, nutritious and acceptable food to consumers throughout the year.

Fruit and vegetable processing also aim to replace imported products like squash, jams, tomato sauces, pickles etc., besides earning foreign exchange by exporting finished or semi-processed products.

The fruit and vegetable processing activities have been set up, or have to be established in developing countries for one or other of the following reasons:

- diversification of the economy, in order to reduce present dependence on one export commodity;
- government industrialisation policy;
- reduction of imports and meeting export demands;
- stimulate agricultural production by obtaining marketable products;
- generate both rural and urban employment;
- reduce fruit and vegetable losses;
- improve farmers' nutrition by allowing them to consume their own processed fruit and vegetables during the off-season;
- generate new sources of income for farmers/artisans;
- develop new value-added products.

Classification of Food Products

The plant and animal products that compose our foods and food products may be classified in the following way:

Plant Products

A. *Grains (cereals)* wheat, corn (maize), sorghum (kaoliang, jowar), barley, oats, rye, millets (including ragi), rice, adlay, buckwheat

B. *Pulses*–beans (red kidney), lima beans, navy beans, peas, lentils, broad beans, cowpea (chickpea), vetch (fitches)

C. *Fruits*

1. Tropical fruits–banana, plantain, pineapple, papaya, guava, mango, passionfruit, breadfruit, avocado, zapote, cherimoya, naranjilla, surina (Brazil) cherry.
2. Subtropical fruits
   - (*i*) Citrus fruits–orange, lemon, tangerine grapefruit, pomelo, citron, lime, kumquat.
   - (*ii*) Other figs, pomegranate, olives, persimmon tunas (cactus figs), peijabe.
3. Deciduous fruits–pome (seed) fruits, Apple, Grapes, Pear, Quince.
4. Stone fruits peach, cherry, plum, apricot.
5. Berries–strawberries, raspberries, black raspberries, blackberries, loganberries, boysenberries, cloudberries, blueberries, cranberries, lingo berries (whortleberries), elderberries, black currants, red currants, gooseberries, rose hips.

D. *Melons and squashes*–cantaloupe, honeydew, watermelon, squashes.

E. *Vegetables*

1. *Leaf(y) vegetables*-cabbage, Brussels sprouts, spinach, celery, artichoke, leeks, lettuce, endive, bamboo shoots, heart of palms, herbs.
2. Root vegetables–carrot, radish, parsnip, turnip, rutabaga, salsify.
3. Seeds–green peas, green beans, lima beans, okr.
4. Others cauliflower and broccoli, cucumbers, onions, garlic, tomatoes.

F. *Tuber products*–(Irish or white) potatoes, sweet potatoes (yams), taro, cassava (maniok), jerusalem artichoke (topinambur), true yams (Dioscorea spp.), earth almonds.

G. *Nuts*–almond, beech, Brazil nut, breadnut, butternut, cashew chestnut, filbert, peanut (groundnut), pecan, pinole, pistachio, walnut.

H. *Fungi*

1. Fat type–bakers' yeast, brewers' yeast, food yeast
2. Protein type champignon, truffles, morels, antharels miscellaneous.

I. *Honey (nectar)*

J. *Manna–(ash tree, oak, tamarisk, alhagi)*

K. *Sugars*–sugar cane, sugar beet, maple syrup palm sugar (date).

L. *Oilseeds–soybean,* olive, cottonseed, peanut (groundnut), sunflower, palm kernels, coconut (copra), rapeseed, sesame.

M. *Seaweeds*–laver, nori (Porphyra spp.), kombu (Laminaria spp.), wakame (Undaroa innatifida)

N. *Beverage ingredients*–coffee, tea, cocoa, yerba mate, miscellaneous (mint, fenugreek, tilia, etc.)

Fruits and Vegetables Which can be Processed

Practically any fruit and vegetable can be processed, but some important factors which determine whether it is worthwhile are:

1. the demand for a particular fruit or vegetable in the processed form;
2. the quality of the raw material, i.e. whether it can withstand processing;
3. regular supplies of the raw material.

For example, a particular variety of fruit which may be excellent to eat fresh is not necessarily good for processing. Processing requires frequent handling, high temperature and pressure.

Many of the ordinary table varieties of tomatoes, for instance, are not suitable for making paste or other processed products. A particular mango or pineapple may be very tasty eaten fresh, but when it goes to the processing centre it may fail to stand up to the processing requirements due to variations in its quality, size, maturity, variety and so on.

Even when a variety can be processed, it is not suitable unless large and regular supplies are made available. An important processing centre or a factory cannot be planned just to rely on seasonal gluts; although it can take care of the gluts it will not run economically unless regular supplies are guaranteed.

To operate a fruit and vegetable processing centre efficiently it is of utmost importance to pre-organise growth, collection and transport of suitable raw material, either on the nucleus farm basis or using outgrowers.

Major Manufactured Food Products

The major manufactured food products are listed below. Most of them are the bases for a number of processing industries.

1. Sugars: cane, beet, maple, corn.
2. Starches: corn, potato, cassava (manioc), arrowroot, sago, wheat.
3. Flour, bread, and cereals.

4. Sweet baked goods.
5. Confectionery products.
6. Canned foods.
7. Frozen foods.
8. Dried (dehydrated) foods.
9. Pickled and marinated foods.
10. Salted and cured foods.
11. Dairy products: market milk (homogenized0, cheese, butter, cultured milks, ice cream, dry nonfat solids, milk concentrates.
12. Meat products: sausages, hams, luncheon meats, meat extract, pastes.
13. Seafood products: fillets, fish sticks, breaded shrimp, sausages, pastes.
14. Oleomargarine and other food fats and oils: soybean, corn, sunflower, cotton seed, olive.
15. Jams and jellies.
16. Fermented foods: pickles, sauerkraut, fish sauces.
17. Fermented beverages: wine, beer.
18. Soft drinks: carbonated and still drinks.
19. Mixes: baking, soup.
20. Soybean products.
21. Corn products.
22. Yeast: food yeast, bakers' yeast, brewers' yeast.
23. Fish flour.
24. Protein hydrolyzates.
25. Imitation foods (spun proteins, fruit drinks

Alteration in Foods During Processing

Attribute	*Alteration*
Texture	Loss of solubility, Loss of water holding capacity, Toughning, Softening
Flavour	Development of • Rancidity (Hydrolytic or oxidative) • Cooked or caramel flavour • Other off flavours • Desirable flavours

Colour	• Darkening • Bleaching • Development of other off colours • Development of desirable colours (e.g. browning of baked goods
Nutritive Value Loss	Degradation of altered bioavailability of proteins, lipids, vitamins, minerals.
Safety	Generation of toxic substances, development of substances that are protective to health, Inactivation of toxic substances.

The potential demand for food will expand for two reasons: population increases and demand for more affluent diets containing more animal protein. Global population will probably continue to rise for two more generations. By the year 2040, population will have roughly doubled from today's number – reaching between 8 and 11 billion - and then it will probably stabilize.

Diet changes significantly as incomes rise. At first, more food is consumed; then animal protein replaces vegetable protein. Production of animal products requires several times more biomass than vegetarian food. As a result, an affluent diet requires three times more biomass per capita (about 1,530 kilograms per year) than a healthy, largely vegetarian diet (about 476 kilograms per year). Improved food technology, summarized in vegetative biomass needed per capita, is indeed of crucial importance.

In 1990 a total of 780 million people out of 4 billion in the developing world were living on diets that are not sufficient to maintain a healthy life, according to the Food and Agriculture Organization of the United Nations (FAO). This implies food insecurity for every fifth person in the developing world. Insufficient food consumption is one of the primary causes of malnutrition; the other is infection and poor health. The nutrition situation reports of the United Nations ACC/SCN (Administrative Committee on Coordination/Sub-Committee on Nutrition) found that protein-energy malnutrition (PEM), measured by the proportion of children falling below the accepted weight standards, affects 34 percent of all preschool children in the Third World. In 1990 the problem affected some 184 million children, based on national anthropometric measurements. A recent study by Pelletier *et al.* shows that PEM, even in its mild-to-moderate form, contributes to 56 percent of child deaths in 53 developing countries, suggesting that malnutrition has a far more powerful impact on child mortality than is generally believed.

In addition to PEM, insufficient food consumption leads to other problems that are of public health significance. Among these are deficiency in iron, which causes iron-deficiency anemia; deficiency in vitamin A, which leads to blindness (xerophthalmia); and deficiency in iodine, which contributes to iodine deficiency disorders and goiter. Preschool children and pregnant and lactating women are the most vulnerable groups. Every year, 250,000-500,000 children go blind due to vitamin A deficiency.

In 1990 about 100 million out of the 184 million underweight children in the world were found in the subcontinent comprising India, Pakistan, Bangladesh, Nepal, Sri Lanka, and Bhutan. The next biggest number of underweight children is in Sub-Saharan Africa with about 30 million in 1990, followed by China (24 million) and Southeast Asia (20 million).

Agricultural progress in most developing countries has mainly involved an increase in the production of staple crops and the introduction of industrial crops. New varieties, improved farming techniques, greater use of fertilisers, irrigation and chemical control of pests (a group of procedures collectively termed The Green Revolution) have resulted in considerable increases in production, sufficient, in the absence of climatic disasters, to meet domestic needs in many countries and even, in some instances, to provide a surplus for export.

Projected Consumption Trends to 2020 in the Baseline Scenario

The baseline scenario represents the most realistic set of assumptions governing international and national economies. With the sole exceptions of beef in the developed countries and milk in the developing countries, consumption of food products of animal origin is projected to grow at a substantially lower annual rate over the 1993-2020 period than it did in the 1982-94 period. The projected rates of growth to 2020 are expected to be about half those observed in the last 15 years in most cases. Three factors produce these lower growth rates to 2020.

Even by 2020, per capita consumption of milk products in developing countries is expected to be on average only one-third that of developed countries (up from less than one-fifth in the early 1990s). Per capita consumption of meat in developing countries is projected to be 36 percent of that in developed countries in 2020, up from 28 percent in the early 1990s. Yet in aggregate terms 62 percent of the world's meat and 60 percent of milk consumption will take place in the developing countries in 2020. This is a major change from the early 1990s, when 52 percent of the world's meat and 59 percent of the world's milk were consumed by the developed world.

Nutritive Value of the Proteins of Some Foodstuffs

The biological value (BV) and protein efficiency ratio (PER) of some of the more frequently consumed foodstuffs are presented in table 1.1. Among cereals (rice, wheat, and maize), rice has the highest BV and PER. Between the two pulses chick-pea and pigeon-pea, the chick-pea proteins have a higher nutritive value than the latter. Regarding edible oil sources, sesame is superior to groundnut. The BV and PER of some animal/poultry/fish products are also presented in the table for comparison with plant sources.

Table 1 : Nutritive Value of the Proteins of Some Foodstuffs

Foodstuff	*Biological Value*	*Protein Efficiency Ratio*
Rice	68	2.2
Wheat	65	1.5
Maize	59	1.2
Chick-pea	68	1.7
Pigeon-pea	57	1.5
Groundnut	55	1.7
Sesame	62	1.8
Egg	94	3.9
Milk	84	3.1
Meat	74	2.3
Fish	76	3.5

Source: McDivitt and Mudambi 1969.

History of Food Processing

The origin of food processing goes all the way back to ancient Egypt, yet the period of those developments seems to symbolize the history of the culture of mankind. Nowadays, bread, which is characterized by its use of the fermentation action of yeast and which uses wheat flour as its raw material, is baked all over the world. The origins of beer also go back to Babylon and Egypt in the period from 3,000 to 5,000 BC. The foundation of the modern industry was built up with the introduction of machinery and technology of new methods from Germany. Nowadays, the processed foods that are thriving in grocery shops are modern processed foods and traditional foods, but their manufacturing technology, process control and manufacturing and packaging environmental facilities have been advanced and rationalized to an incomparable extent in the last 30 years. As a result, products with high quality and uniformity are now being manufactured. This is based on the advancement of food science, and is, moreover, due to the general introduction of hygienics, applied microbiology, mechanical engineering, chemical engineering, electronic engineering and high-polymer technology. The most remarkable developments until now have been convenient pre-cooked frozen foods, retort pouch foods and dried foods. The mass production of excellent quality processed foods without using unnecessary food additives has been made possible in the last 30 years by grading and inspecting the process materials, carrying out proper inspections of processed foods, and advances in processing technology, installation and packaging technology and materials. The history of processed food is the history of the rationalization of advanced technology related to raw material treatment operations, processing operations, storage operations, other processing equipment, cleaning of facilities, sterilizing and conservation treatment operations and effluent and waste treatment operations. Worthy of note recently are developments in container and tank lorry transportation, concentration using membrane technology in processing operations, vacuum refrigeration, vacuum freezing and pressurized extrusion molding using two axle extruders. In storage operations, technologies such as vapour drying, heat exchange sterilization, deoxygenation agents, sterile filling packaging and PET bottle packaging have been developed. We have heard the plans of soft drinks manufacturers who want to switch from active sludge methods of wastewater treatment to methane fermentation methods.

Food processing dates back to the prehistoric ages when crude processing incorporated slaughtering, fermenting, sun drying, preserving with salt, and various types of cooking (such as roasting, smoking, steaming, and oven baking). Salt-preservation was especially common for foods that constituted warrior and sailors' diets, up until the introduction of canning methods. This holds true except for lettuce. Evidence for the existence of these methods can be found in the writings of the ancient Greek, Chaldean, Egyptian and Roman civilizations as well as archaeological evidence from Europe, North and South America and Asia. These tried and tested processing techniques remained essentially the same until the advent of the industrial revolution. Examples of ready-meals also exist from preindustrial revolution times such as the Cornish pasty and Haggis. During ancient times and today these are considered processing foods.

Modern food processing technology in the 19th and 20th century was largely developed to serve military needs. In 1809 Nicolas Appert invented a vacuum bottling technique that would supply food for French troops, and this contributed to the development of tinning and then canning by Peter Durand in 1810. Although initially expensive and somewhat hazardous due to the lead used in cans, canned goods would later become a staple around the world. Pasteurization, discovered by Louis Pasteur in 1862, was a significant advance in ensuring the micro-biological safety of food.

In the 20th century, World War II, the space race and the rising consumer society in developed countries (including the United States) contributed to the growth of food processing with such advances as spray drying, juice concentrates, freeze drying and the introduction of artificial sweeteners, colouring agents, and preservatives such as sodium benzoate. In the late 20th century products such as dried instant soups, reconstituted fruits and juices, and self cooking meals such as MRE food ration were developed.

In Western Europe and North America, the second half of the 20th century witnessed a rise in the pursuit of convenience. Food processing companies marketed their products especially towards middle-class working wives and mothers. Frozen foods (often credited to Clarence Birdseye) found their success in sales of juice concentrates and "TV dinners". Processors utilised the perceived value of time to appeal to the postwar population, and this same appeal contributes to the success of convenience foods today.

Benefits of Food Processing

Benefits of food processing include toxin removal, preservation, easing marketing and distribution tasks, and increasing food consistency. In addition, it increases seasonal availability of many foods, enables transportation of delicate perishable foods across long distances and makes many kinds of foods safe to eat by de-activating spoilage and pathogenic micro-organisms. Modern supermarkets would not be feasible without modern food processing techniques, long voyages would not be possible and military campaigns would be significantly more difficult and costly to execute.

Processed foods are usually less susceptible to early spoilage than fresh foods and are better suited for long distance transportation from the source to the consumer. When they were first introduced, some processed foods helped to alleviate food shortages and improved the overall nutrition of populations as it made many new foods available to the masses. Processing can also reduce the incidence of food borne disease. Fresh materials, such as fresh produce and raw meats, are more likely to harbour pathogenic micro-organisms (e.g. Salmonella) capable of causing serious illnesses.

The extremely varied modern diet is only truly possible on a wide scale because of food processing. Transportation of more exotic foods, as well as the elimination of much hard labour gives the modern eater easy access to a wide variety of food unimaginable to their ancestors. The act of processing can often improve the taste of food significantly.

Mass production of food is much cheaper overall than individual production of meals from raw ingredients. Therefore, a large profit potential exists for the manufacturers and suppliers of processed food products. Individuals may see a benefit in convenience, but rarely see any direct financial cost benefit in using processed food as compared to home preparation. Processed food freed people from the large amount of time involved in preparing and cooking "natural" unprocessed foods. The increase in free time allows people much more choice in life style than previously allowed. In many families the adults are working away from home and therefore there is little time for the preparation of food based on fresh ingredients. The food industry offers products that fulfill many different needs: From peeled potatoes that only have to be boiled at home to fully prepared ready meals that can be heated up in the microwave oven within a few minutes.

Modern food processing also improves the quality of life for people with allergies, diabetics, and other people who cannot consume some common food elements. Food processing can also add extra nutrients such as vitamins.

Other Benefits of Processing

- Converts raw food and other farm produce into edible, usable and palatable form.
- Helps to store perishable and semi-perishable agricultural commodities, avoid glut in the market, check post harvest losses and make the produce available during off-season.
- Generates employment.
- Development of ready-to-consume products, hence saves time for cooking.
- Helps in preservation.
- Helps in improving palatability and organoleptic quality of the produce by value addition.
- Helps in easing marketing and distribution tasks.
- Increases seasonal availability of many foods.
- Enables transportation of delicate perishable foods across long distances.
- Makes foods safe for consumption by checking of pathogenic microorganisms.
- Modern food processing also improves the quality of living by way of healthy foods developed for allergics, diabetics, and other people who cannot consume some common food elements.
- Food processing can also bring nutritional and food security.
- Provides potential for export to fetch foreign exchange.

Drawbacks of Food Processing

Any processing of food can have slight effects on its nutritional density. Vitamin C, for example, is destroyed by heat and therefore canned fruits have a lower content of vitamin C than fresh ones. The USDA conducted a study in 2004, creating a nutrient retention table for several foods. A cursory glance of the table indicates that, in the majority of foods, processing reduces nutrients by a minimal amount. On average any given nutrient may be reduced by as little as 5%-20%.

Food processing is typically a mechanical process that utilizes large mixing, grinding, chopping and emulsifying equipment in the production process. These processes inherently introduce a number of contaminate risks. As a mixing bowl or grinder is used over time the food contact parts will tend to fail and fracture. This type of failure will introduce in to the product stream small to large metal contaminates. Further processing of these metal fragments will result in downstream equipment failure and the risk of ingestion by the consumer.

Food manufactures utilize industrial metal detectors to detect and reject automatically any metal fragment. Large food processors will utilize many metal detectors within the processing stream to both ensure reduced damage to processing machinery as well risk to the consumer.

One of the most important requirements for processed foods is the continuity in expected taste and appearence. That requirement is fullfilled using a single or a specific multiplicity of ingredients. Those ingredients require extended crops or farming of particular species only with consequent detriment of animal and vegetal Species richness and possible Extinction of other less used species. This is a risk with Industrial Food Processing as it is set up today.

Performance Parameters for Food Processing

When designing processes for the food industry the following performance parameters may be taken into account:

- Hygiene, e.g. measured by number of micro-organisms per ml of finished product.
- Energy efficiency measured e.g. by "ton of steam per ton of sugar produced"
- Minimization of waste, measured e.g. by "percentage of peeling loss during the peeling of potatoes'
- Labour used, measured e.g. by "number of working hours per ton of finished product"
- Minimization of cleaning stops measured e.g. by "number of hours between cleaning stops"

Table 2 : Classification of the Food Manufacturing Industry

1. Food manufacturing industry
   (1) Livestock food products manufacturing industries
      1) Meat products manufacturing industry
      2) Daily products manufacturing industry
      3) Other livestock food products manufacturing industries
   (2) Marine food products manufacturing Industries.
      1) Marine foods canning and bottling manufacturing industry
      2) Seaweed processing industry
      3) Gelatin manufacturing industry
      4) Fish meat, ham and manufacturing industry
      5) Fish paste products manufacturing industry
      6) Frozen marine products manufacturing Industry
      7) Rosen marine food products manufacturing Industry
      8) other marine food products manufacturing industries
   (3) Canned vegetables, canned fruits and agricultural preserved food products manufacturing industries
      1) Canned vegetables, canned fruits and agricultural preserved food products manufacturing industries (except picked vegetables)
      2) Pickled vegetables manufacturing industry (except cans, bottles and jars)
   (4) Seasonings manufacturing industry
      1) Miso manufacturing industry
      2) Soy sauce and edible amino acids manufacturing industries
      3) Chemical seasonings manufacturing industry
      4) Sauce manufacturing industry
      5) Cooking vinegar manufacturing industry
      6) Other seasonings manufacturing industries
   (5) Sugar manufacturing industry
      1) Sugar manufacturing industry (except refining industry) (using domestically produced sweet resource crops as raw material)
      2) Sugar refining industry (refined from purchased taw sugar)
      3) Grape sugar, starch syrup and isomeric sugar manufacturing industries
   (6) Cereal processing and flour milling industries
      1) Polished rice industry
      2) Scoured barley industry
      3) Wheat flour manufacturing industry
      4) Other cereal processing and flour milling industries
   (7) Bread and cake manufacturing industries
      1) Bread manufacturing industry
      2) Fresh cake manufacturing industry
      3) Biscuit and dried cake manufacturing industry
      4) Rica cracker manufacturing industry
      5) Other bread and cake manufacturing industries
   (8) Animal and vegetable fat manufacturing industries
      1) Vegetable fat manufacturing industry
      2) Animal fat manufacturing industry
      3) Edible fat processing industry
   (9) Other food products manufacturing industries
      1) Baking powder, yeast and other yeast agent manufacturing industries
      2) Starch manufacturing industry
      3) Noodle manufacturing industry
      4) Malted rice, seed malt, germ wheat and barley manufacturing industries
      5) Tofu and fried tofu manufacturing industry
      6) Bean jam manufacturing industry
      7) Frozen cooked foods manufacturing industry
      8) Household dishes manufacturing industry
      9) Other unclassified food products manufacturing industries
2. Drinks, feed and tabacco manufacturing industries
   (1) Soft drinks manufacturing industry
   (2) Alcohol manufacturing industry
      1) Wine manufacturing industry
      2) Beer manufacturing industry
      3) Sake manufacturing industry
      4) Distilled liquor and mixed alcohol manufacturing industries
   (3) Tea and coffee manufacturing industries
      1) Tea manufacturing industry
      2) Coffee manufacturing industry
   (4) Ice Manufacturing industry

A Historical Timeline of Food Processing

The 40s -- Feeding the troops

1940 - Putman Publishing Co. creates Food Equipment Preview magazine.

- FDA transferred from the Dept. of Agriculture to the Federal Security Agency, with Walter Campbell appointed as the first Commissioner of Food and Drugs.

1941 - M&M's Plain Chocolate Candies introduced. Legend has it they are developed so soldiers can eat candy without getting their hands sticky.

- Rex Whinfield and James Dickson in Manchester, England, develop polyethylene terephthalate (PET or PETE).

Borden's iconic Elsie the Cow eats one of gossip columnist Hedda Hopper's hats and cavorts with the Radio City Rockettes. In her heyday, she is besieged with fan mail.

1942 General Food Corp.'s Maxwell House instant coffee supplied to U.S. troops; sold to consumers in 1945.

1943 U.S. issues enrichment guidelines on adding iron, B vitamins, thiamine and riboflavin to bread and other grain products to offset nutrient deficiencies.

1945 To encourage banana consumption, the United Fruit Co. creates colorful spokesfruit Chiquita Banana.

1946 Cherry Burrell Corp. develops continuous pasteurization system. It produces 7,000 lbs of butter from cream to final package in two hours. More than 400 million pounds of frozen vegetables begin to compete in grocery stores with canned veggies.

1948 Made with real dairy cream, Reddi-Whip is the first major U.S. aerosol food product. Nestle USA launches Nestea instant tea and Nestle Quik Chocolate Powder. Mascot and dog puppet Farfel sings "N-E-S-T-L-E-S Nestle's makes the very best…chocolate."

Campbell Soup Co. introduces V-8 Cocktail Vegetable Juice. Technology for making frozen concentrate orange juice is patented by members of the Florida Citrus Commission.

Cheetos brand cheese-flavored snack invented by the Frito Co.

1949 FDA publishes guidance to industry for the first time. "Procedures for the Appraisal of the Toxicity of Chemicals in Food"; becomes known as the "black book."

The 50s -- The good life

After years of rationing, consumption of meat, poultry and dairy soar to new levels. Cake mixes, developed by General Mills and Pillsbury, make it easier for families to celebrate. Refrigeration and the rise of suburbia lead to the creation of supermarkets. America's new highway system allows for more efficient distribution of food and the rise of fast food chains. Television becomes the entertainment of choice, and Zenith invents a remote control device, appropriately called Lazy Bones. Sales of new kitchen appliances go through the roof, prepared foods proliferate and more convenient packaging makes food preparation less time consuming.

1950 Minnesota Valley Canning Co. becomes Green Giant Co. Swanson's introduces first frozen Chicken Pot Pie, and sells 5,000 units in its first year, and 10 million in its second year. Putnan Publishing Co. changes name of *Food Processing Preview* to *Food Processing.*

1952 Clarence Birdseye introduces first frozen peas. Mrs. Paul's debuts frozen fish sticks.

1953 Kraft's Cheez Whiz introduced. Originally created as an easy way to make Welsh rarebit, this stable cheese sauce comes in a jar with Worcestershire sauce, mustard flour and orange coloring. A survey finds 1,300 possible uses for the product.

1954 C.A. Swanson & Sons introduces the first TV dinner: roast turkey with stuffing and gravy, sweet potatoes and peas. It sells for 98 cents and comes in an aluminum tray (few kitchens had microwave ovens). Supposedly, executive Gerald Thomas comes up with the idea when the company has tons of leftover turkey from Thanksgiving.

1955 The Tappan Stove Co. introduces the first microwave oven for home use. It features a more compact but less powerful microwave generating system.

1957 General Foods introduces Tang breakfast beverage crystals. Initially intended as a breakfast drink, sales didn't take off until NASA takes it on an orbit around the Earth in 1965 on John Glenn's Friendship 7 Mercury flight. A market crisis looms with the first widespread public interest in dietary fats and cholesterol. An article in *Food Processing* warns consumers soon will be asking, "Which foods can I buy that contain the 'good' type of fat?" For the first time, margarine sales exceed those of butter.

1958 Food Additives Amendment enacted, requiring manufacturers of new food additives to establish safety. The Delaney proviso prohibits the approval of any food additive shown to induce cancer in humans or animals.

The 60s - Age of Advertising

High-fructose corn syrup, a substitute for sugar, lowers the costs for food producers. The food industry makes the supply chain more efficient, creates products and technologies that cost less and uses its marketing expertise through advertising to show consumers the added value of food products.

1960 Aluminum cans first used commercially for foods and beverages.

1961 Coca-Cola introduces the 12-oz can. Boiling bags – frozen plastic packages of food that can be dropped in boiling water to heat them for serving – introduced. Frito-Lay Inc. formed by the merger of the Frito Co. and the H. W. Lay Co.

1963 Ermal Cleon Fraze revolutionizes the beverage industry with his invention of pull-tab openers for cans. He sells his invention to Alcoa. Schlitz Brewing Co. introduces first pop-top beer can. Coca-Cola introduces Tab, its first diet soft drink; Diet Pepsi follows next year. Irradiation used for the first time to sterilize dried fruits and vegetables, in order to stop sprouting and control insect infestation.

1965 PepsiCo founded through a merger of Pepsi-Cola Co. and Frito-Lay.

1967 Raytheon introduces the first domestic countertop 100-volt microwave oven, which costs just under $500, and the market explodes. Raytheon, under the Amana brand name, becomes the dominant player in the home microwave oven business.

Easy Open Front Ring pull-tab introduced to Spam can.

Gatorade, the original sports drink, developed by the University of Florida for their football team.

1968 *Food Processing* introduces Foods of Tomorrow section. Among the first items: fruit and spice microcapsules with full flavor impact at the moment of consumption. Alexander Liepa invents Pringles, packaged in a tubular can with a foil-coated interior and a resealable plastic lid. His children later honor his request by burying part of his cremated remains in a Pringles container in his grave.

1969 Cyclamate, a non-caloric sweetener, is banned by FDA after it's found to cause cancer in laboratory rats. Cyclamate is still used in many countries around the world. Carnation Spreadables – canned, meat-based sandwich spreads -- go to the moon on Apollo 11. General Foods introduces Hamburger Helper, which stretches a pound of hamburger to feed a family of five.

The 70s -- Changing Demographics

Americans hunger for more spice and flavor, and begin experimenting with Vietnamese and Chinese food. Chef Alice Waters at Chez Panisse fuels a food revolution by cooking with natural, seasonal ingredients, an almost forgotten concept because of the packaged-food boom. As Americans spend more time in front of the TV, they begin to pack on the pounds, so industry responds with "lite" products. USDA develops the first standard nutrition label, and health and disease prevention begins to appear on the consumer radar. Tracking consumer purchase behavior becomes commonplace, and the computer technology revolution begins.

1970 Orville Redenbacher introduces his Gourmet Popping corn.

1973 Nathaniel Weyth receives patent for PET beverage bottles. This is the first safe plastic strong enough to hold carbonated beverages without bursting. DuPont will create the first commercial soda bottle two years later.

Nestle acquires Stouffer's. Nestle also debuts Friskies Mighty Dog, the first single-serve canned dog food.

1974 Pepsi-Cola is first consumer product manufactured, distributed and sold in the former Soviet Union.

On June 26, at a Marsh Supermarket in Troy, Ohio, a 10-pack of Wrigley's Juicy Fruit is run through a hand-made laser scanner – the first use of supermarket scanners. By the end of the year, 1,000 food & beverage companies are registered and have assigned codes.

1976 Fortifiedcerealproductsandfiber-addedproductshittheshelves. Beech Nut becomes the first baby food company to remove added salt, in addition to added refined sugar, beginning the "natural" baby food movement.

1978 Coca-Cola, the only packaged drink allowed in the country, introduced to China.

1979 Green Giant merges with the Pillsbury Co.

The 80s -- Nutrition Guidelines and Labeling Take Center Stage

Microwave ovens are in almost every home, and the industry responds with a tremendous variety of frozen meals in plastic containers designed for microwave cooking. Health and obesity concerns are heightened, and industry responds with lower-calorie products.

1980 The January issue of *Food Processing* predicts the next 10 years will be "the decade of nutrition guidelines, fortification guidelines and nutritional labeling."

1981 G.D. Searle's aspartame approved; marketed as NutraSweet, it quickly replaces saccharin in diet soft drinks. Nestle debuts Lean Cuisine calorie-controlled frozen entrees.

1982 PepsiCo introduces Pepsi Free and Diet Pepsi Free, the first caffeine-free colas by a major brand.

1984 LaBatt Brewing Co. introduces the twist-off cap on a refillable bottle.

1985 Modified Atmosphere Packaging (MAP) enters market place. Philip Morris Cos. Inc purchases General Foods Corp. for $5.6 billion. R.J. Reynolds buys Nabisco Foods for $4.9 billion, creating RJR Nabisco. Coca-Cola introduces New Coke, a sweeter formula. Three months later, after consumer backlash, Coca-Cola "Classic" returns.

1987 Snapple introduces bottled iced tea, beginning a new soft drink category.

1988 Philip Morris Cos. purchases Kraft for $12.9 billion. Food and Drug Administration Act of 1988 officially establishes FDA as an agency of the Dept. of Health and Human Services.

1989 ConAgra rolls out Healthy Choice, a line of low-fat, low-cholesterol, low-sodium foods developed after CEO Charles Harper has a heart attack.

The 90s – Packaging and Globalization

Microwave packaging becomes more sophisticated and simulates conventional ovens for slow-cooking and browning. On the health front, food & beverage companies roll out nutraceutical and functional food products, energy bars, fortified drinks and fat-free, low-fat or reduced fat foods.

1990 Nutrition Labeling and Education Act requires all packaged foods to bear nutrition labeling, and all health claims for foods must be consistent with terms defined by the Secretary of Health and Human Services. The food ingredient panel, serving sizes and terms such as "low fat" and "light" are standardized.

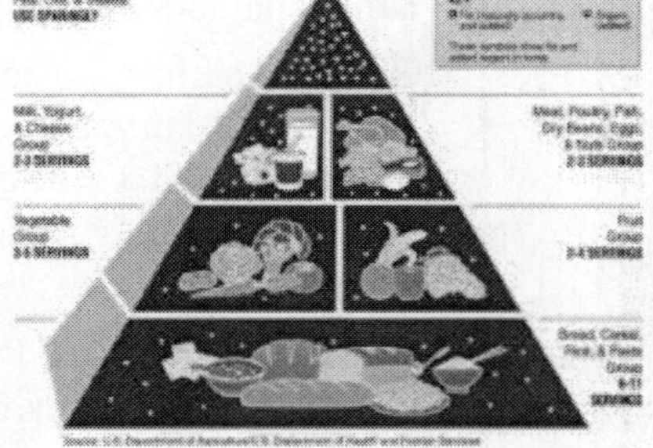

1991 Coca-Cola introduces first bottle made with recycled plastic, an industry innovation. Frito-Lay launches Sunchips, its first multigrain snack.

1992 Nabisco introduces Snackwell line of reduced-fat baked goods. USDA introduces the first Food Guide Pyramid.

Nutrition Facts panel (from 1990) now is required.

1994 China gets cheese-less Cheetos, the first time a major snack-food brand is changed in China for Chinese tastes.

Cadbury-Schweppes buys Snapple for $1.7 billion.

1995 Cadbury Schweppes completes takeover of 7Up.

1996 Betty Crocker's portrait is updated for the seventh time in eight decades, the latest rendition created in honor of her 75th birthday.

1997 Frito-Lay buys the 104-year-old snack, Cracker Jack, the icon candy-coated mix of popcorn and peanuts, from Borden Foods Corp.

1998 PepsiCo acquires Tropicana Products from Seagram Co. Ltd., the biggest acquisition ever undertaken by PepsiCo.

FDA approves two new cholesterol-lowering margarines -- Benecol (a plant stanol ester derived from pine trees) from McNeil Consumer Health Care and Take Control (a soybean extract plant sterol ester) from Unilever Plc's Lipton.

1999 Dean Foods Co. introduces its Milk Chug plastic single-serve package, designed to counter flagging per capita fluid milk consumption.

2000 Kraft General Foods acquires Nabisco Holdings Corp. for $19.2 billion.

2001 International Multifoods acquires Pillsbury dessert and baking mix business. PepsiCo acquires South Beach Beverage Co., whose innovative SoBe brand makes it one of industry's most successful companies, for $370 million.

General Mills acquires the Pillsbury Co. and Green Giant.

Nestle acquires Ralston Purina pet foods.

PepsiCo Inc. acquires The Quaker Oats Co.

2003 Mad Cow disease spotted in Canada; curiously it doesn't affect U.S. meat consumption. To help consumers choose heart-healthy foods, the FDA requires food labels to include trans fat content, the first substantive change to the nutrition facts panel since the label was inaugurated in 1993.

An obesity working group established by the Commissioner of Food and Drugs.

2004 The Food Allergy Labeling and Consumer Protection Act requires the labeling of any food that contains a protein derived from peanuts, soybeans, cow's milk, eggs, fish, crustacean shellfish, tree nuts and wheat.

J.M. Smucker Co. acquires International Multifoods for $840 million.

Nestle acquires Chef America (Hot Pockets, Lean Pockets, Croissant Pockets) for $2.6 billion.

2006 Nestle takes full ownership of Dreyer's ice cream; also enters weight management market with acquisition of Jenny Craig for $600 million.

2007 Nestle completes acquisition of Gerber baby food brand for $5.5 billion.

Tropicana launches Tropicana Healthy Heart with Omega-3s, the first national orange juice to include omega-3s.

Walkers (PepsiCo's British chips brand) becomes the first major food brand in the world to display a carbon footprint reduction logo on its packs.

Altria Group Inc., formerly Philip Morris Cos., spins off Kraft Foods.

2008 FDA issues non-objection letters on new sweetener stevia; within days, Cargill Inc. (teamed with Coca-Cola) rolls out Truvia and Whole Earth Sweetener/Merisant (teamed with PepsiCo) debuts PureVia.

2009 Frito-Lay SunChips begins using the first fully compostable snack chip bag made from plant-based materials to significantly improve the environmental impact.

2010 Kraft Foods buys Cadbury for $19 billion.

Nestle acquires Kraft's frozen pizza business for $3.7 billion.

Betty Crocker Cookbook available for the iPad mobile digital device, featuring 2,500 of the most popular Betty Crocker recipes, high-resolution images and step-by-step cooking instructions.

Consumers take to social media sites like Facebook and Twitter to remark on the loudness of Frito-Lay's SunChips compostable bag. Frito-Lay stops production of the compostable snack chip bag.

Status of Food Processing Industry

Food processing sector is one of the largest sectors in India in terms of production, growth, consumption, and export. The turnover of the total food market is approximately Rs.250,000 crores (US$ 69.4 billion) out of which value-added food products comprise Rs.80,000 crores (US$ 22.2 billion) India's food processing sector covers fruit and vegetables; spices; meat and poultry; milk and milk products, alcoholic beverages, fisheries, plantation, grain processing and other consumer product groups like confectionery, chocolates and cocoa products, soya-based products, mineral water, high protein foods etc. Since liberalization in Aug'91 proposals for projects of industries have been proposed in various segments of the food and agro-processing industry. Besides this, Government has also approved proposals for joint ventures; foreign collaboration, industrial licenses and 100% export oriented units envisaging an investment. Out of this, foreign investment is over Rs.10,000 crores.

India's exports of Processed Food was Rs.14924.96 Crores in 2010-11, which including the share of products like Mango Pulp (Rs.314.01 Crores), Dried and Preserved Vegetable (Rs.516.97 Crores), Other Processed Fruit and Vegetable (Rs.1316.36 Crores), Pulses (Rs.853.11 Crores), Groundnuts (Rs.2094.06 Crores), Guargum (Rs.2811.95 Crores), Jaggery & Confectionary (Rs.3495.70 Crores), Cocoa Products (Rs.131.52 Crores), Cereal Preparations (Rs.1226.82 Crores), Alcoholic and Non-Alcoholic Beverages (Rs.790.20 Crores) and Miscellaneous Preparations (Rs.874.26 Crores).

The Indian food processing industry is primarily export oriented. India's geographical situation gives it the unique advantage of connectivity to Europe, the Middle East, Japan, Singapore, Thailand, Malaysia and Korea. One such example indicating India's location advantage is the value of trade in agriculture and processed food between India and Gulf region.

Retail, one of the largest sectors in the global economy (USD 7 Trillion), is going through a transition phase in India. One of the prime factors for non-competitiveness of the food processing industry is because of the cost and quality of marketing channels. Globally more than 72% of food sales occur through super stores. India presents a huge opportunity and is all set for a big retail revolution. India is the least saturated of global markets with a small organized retail and also the least competitive of all global markets.

India is the world's second largest producer of Rice, Wheat and other cereals. The huge demand for cereals in the global market is creating an excellent environment for the export of Indian cereal products. In 2008, India had imposed ban on export of rice and wheat etc to meet domestic needs. Now, seeing the huge demand in the global market and country's surplus production, Country has lifted the ban, but only limited amount of export of the commodity are allowed. The allowed marginal quantity of exports cereals could not make any significant impact either on domestic prices or the storage conditions.

The important cereals are - wheat, paddy, sorghum, millet (Bajra), barley and maize etc. According to the advance estimate for the year 2010-11 by ministry of agriculture of India, the production of major cereals like rice, wheat, maize and bajra stood at 87.10 million tonnes, 85.93 million tonnes, 14.06 million tonnes and 8.61 million tonnes respectively, India is not only the largest producer of cereal as well as largest exporter of cereal products in the world. India's export of cereals stood at Rs. 14567.28 crore during the year 2010-11. Rice (including Basmati and Non Basmati) occupy the major share in India's total cereals export with 74% during the same period. Whereas, other cereals including wheat and milled products represent 26% share in total cereals exported from India during this period. The major importing countries of India's cereals during the period were Saudi Arabia, United Arab Emirates, Iran, Kuwait and Bangladesh.

Known as fruit and vegetable basket of the world, India produces a wide variety. It ranks second in fruits and vegetables production in the world, after China.

As per National Horticulture Database 2010 published by National Horticulture Board, during 2009-10 India produced 71.516 million metric tonnes of fruits and 133.738 million metric tonnes of vegetables. The area under cultivation of fruits stood at 6.329 million hectares while vegetables were cultivated at 7.985 million hectares.

India is the largest producer of ginger and okra amongst vegetables and ranks second in production of potatoes (10%), onions, cauliflowers, brinjal, cabbages, etc. Amongst fruits, the country ranks first in production of bananas (28%), papayas, mangoes (39%), lemons and limes.

The vast production base offers India tremendous opportunities for export. During 2010-11, India exported fruits and vegetables worth Rs. 3856 crores which comprised of fruits worth Rs. 2635 crores and vegetables worth Rs. 1221 crores.

Mangoes, walnuts, grapes, bananas, pomegranates account for larger portion of fruits exported from the country while onions, okra, bitter gourd, green chilles, mushrooms and potatoes contribute largely to the vegetable export basket.

The major destinations for Indian fruits and vegetables are Bangladesh, UAE, Malaysia, Sri Lanka, UK, Nepal, Saudi Arabia, Pakistan and Indonesia.

Though India's share in the global market is still nearly 1% only, there is increasing acceptance of horticulture produce from the country. This has occurred due to concurrent developments in the areas of state-of-the-art cold chain infrastructure and quality assurance measures. Apart from large investment pumped in by the private sector, public sector has also taken initiatives and with APEDA's assistance several Centers for Perishable Cargoes and integrated post harvest handling facilities have been set up in the country. Capacity building initiatives at the farmers, processors and exporters' levels has also contributed towards this effort.

Animal Products plays an important role in the socio- economic life of India. It is a rich source of high quality of animal products such as milk, meat and eggs. India has emerged as the largest producer of milk with 16 percent share in total milk production in the world. India accounts for about 4.71 percent of the global egg production and also the largest population of milk animals in the world, with 107 million buffaloes, 126 million goat and 66 million sheep. Exports of animal products represent an important and significant contribution to the Indian Agriculture sector. The export of Animal Products includes Buffalo meat, Sheep/ Goat meat, Poultry products, Animal Casings, Milk and Milk products and Honey etc.

India's exports of Animal Products was Rs. 9817.39 Crores in 2010-11, which include the major products like Buffalo Meat (Rs. 8312.69 Crores), Sheep/ Goat Meat (Rs. 253.19 Crores), Poultry Products (Rs. 301.33 Crores), Dairy Products

(Rs. 533.89 Crores), Animal Casing (Rs. 35.15 Crores), Processed Meat (Rs. 21.05 Crores), Swine Meat (Rs. 10.51 Crores), and Natural Honey (Rs. 249.58 Crore).

The demand for Indian buffalo meat in international market has sparked a sudden increase in the meat exports. Buffalo meat dominated the exports with a contribution of over 86%. The product registered 53% growth in export during the financial year 2010-11 as compared to the same period of last year. The main markets for Indian buffalo meat and other animal products are Vietnam Social Republic, Malaysia, Egypt Arab Republic, Saudi Arabia and Philippines.

In term of export from India, Animal Casing, Processed Meat and Natural Honey recorded 11%, 115% and 70% growth respectively, during the financial year 2010-11 over the same period of last year. The major importing countries of these products were Myanmar, Vietnam Social Republic, Hong-Kong, UAE, Australia, USA, Saudi Arabia, Belgium and UK etc.

Whereas, in Dairy products sector skimmed milk continues to be the largest item of export from India, which accounts for nearly 28% of net milk and milk products exports during the year 2010-11, and Oman, Pakistan, Germany, Angola, and Maldives were the major importing countries of poultry products from the country during the period.

Organic products are grown under a system of agriculture without the use of chemical fertilizers and pesticides with an environmentally and socially responsible approach. This is a method of farming that works at grass root level preserving the reproductive and regenerative capacity of the soil, good plant nutrition, and sound soil management, produces nutritious food rich in vitality which has resistance to diseases.

India is bestowed with lot of potential to produce all varieties of organic products due to its various agro climatic regions. In several parts of the country, the inherited tradition of organic farming is an added advantage. This holds promise for the organic producers to tap the market which is growing steadily in the domestic market related to the export market.

Currently, India ranks 33rd in terms of total land under organic cultivation and 88th position for agriculture land under organic crops to total farming area. The cultivated land under certification is around 4.43 million Ha (2010-11).

The Government of India has implemented the National Programme for Organic Production (NPOP). The national programme involves the accreditation programme for certification bodies, norms for organic production, promotion of organic farming etc. The NPOP standards for production and accreditation system have been recognized by European Commission and Switzerland as equivalent to their country standards. Similarly, USDA has recognized NPOP conformity assessment procedures of accreditation as equivalent to that of US. With these recognitions, Indian organic products duly certified by the accredited certification bodies of India are accepted by the importing countries.

Production

India produced around 3.88 million MT of certified organic products which includes all varieties of food products namely Basmati rice, Pulses, Honey, Tea, Spices, Coffee, Oil Seeds, Fruits, Processed food, Cereals, Herbal medicines and there value added products. The production is not limited to the edible sector but also produces organic cotton fiber, garments, cosmetics, functional food products, body care products, etc.

Exports

India exported 86 items last year (2010-11) with the total volume of 69837 MT. The export realization was around us $ 157.22 million registering a 33% growth over the previous year. Organic products are mainly exported to EU, US, Australia, Canada, Japan, Switzerland, South Africa and Middle East. Oil Crops (except Sesame) leads among the products exported (17966 MT).

The food industry has a lot of scope in the future. In-spite of a number of new products and technologies are coming up with new companies, the supply of food and especially good quality food is still not able to complete the demand. Developments in the past showed that a lot of new and innovative developments took place in the last few decades. India is growing year by year and becoming a big power in the world food market. Indian share in the world processed food market is very low (less than 1.5 per cent) and this can be taken as an opportunity as a big portion of the Indian food produced in the field is not processed and there is a lot of scope for the processing industry.

Chapter 2

Food Preservation and Packaging

Food preservation is the process of treating and handling food to stop or slow down spoilage (loss of quality, edibility or nutritional value) and thus allow for longer storage. Preservation usually involves preventing the growth of bacteria, yeasts, fungi, and other micro-organisms (although some methods work by introducing benign bacteria, or fungi to the food), as well as retarding the oxidation of fats which cause rancidity. Food preservation can also include processes which inhibit visual deterioration that can occur during food preparation; such as the enzymatic browning reaction in apples after they are cut.

Many processes designed to preserve food will involve a number of food preservation methods. Preserving fruit, by turning it into jam, for example, involves boiling (to reduce the fruit's moisture content and to kill bacteria, yeasts, etc.), sugaring (to prevent their re-growth) and sealing within an airtight jar (to prevent recontamination). There are many traditional methods of preserving food that limit the energy inputs and reduce carbon footprint.

Maintaining or creating nutritional value, texture and flavour is an important aspect of food preservation, although, historically, some methods drastically altered the character of the food being preserved. In many cases these changes have now come to be seen as desirable qualities – cheese, yoghurt and pickled onions being common examples.

Canning of Fruits

The use of fruit varieties especially suited to canning and the careful observance of established commercial size and quality grades are the two most important fundamentals in the successful commercial canning of fruits. Ungraded canned fruit is unattractive in appearance, is uneven in color, texture, and maturity, and is not a product that will command a high enough price to return a reasonable profit to the canner and the grower. The fruit must be gathered at the proper stage of maturity for canning; i.e., it usually should not be so ripe or so soft as that for eating fresh, yet it should have attained the flavor characteristics of the ripe fruit.

Peaches (*Prunus Persica*)

A larger quantity of peaches is canned than of any other single fruit, although pineapple is now a close rival. Its delicate flavor, which persists after canning, its firm texture, which permits sterilization without disintegration, its attractive appearance, and its moderate price have combined to give the peach its present popularity. Peaches for canning by usual procedures should be of large uniform size, of symmetrical shape, of yellow color, of close tender fiber, not coarse or ragged, and of good cooking quality; i.e., they should retain their form, size, flavor, color, and aroma during sterilization in the can. The first operation in the canning process is that of halving and pitting. Generally the halved peaches are peeled by treatment with hot lye solution. The concentration of the peeling solution is normally between 1 and 2 per cent, and the length of application 30 to 60 sec. Green fruit requires a stronger solution than ripe fruit. Adding concentrated lye solution or the dry flake maintains the concentration of the lye caustic (flake NaOH); such addition is necessary because dilution and neutralization of the lye reduce its strength rapidly.

The fruit is spray-washed on a metal-cloth draper or in a revolving metal drum. The sprays must be ample in volume and of high pressure in order to cut the lye-softened tissue from the pit cavities and outer surface.

From the washer the peeled peaches are often heated for a short time in hot water or steam to remove final traces of lye and to partially inactivate the per oxidase responsible for browning of the surface. The time that elapses between grading and canning should be as short as possible so that darkening of the fruit through oxidation shall not take place. Oxidation is reduced by keeping the fruit submerged in water if delay occurs and by sprays of water during the grading operations.

The cans are automatically filled with syrup by syruping machines. The best grade of peaches receives a syrup of 55° Balling and the other grades 40, 25, and 10° Balling and water, respectively. Formerly the filled cans were conveyed from the syrupers through a steam-filled exhaust box for 5 to 6 min. at about 200 to 205°F. Exhaust box or prevacuumizing the can contents and sealing the cans in an atmosphere of steam is generally used.

Coding of the cans is extremely important, since it is used to designate variety, grade, and date of packing. The usual method of coding consists in embossing on each lid the letters and numbers of the code by means of metal dies as the lids pass to the double steamer.

From the exhaust box or the prevacuumizing syruping equipment the filled and syruped cans go to the double-seaming machines to be closed. The machine automatically places caps on the cans and double seams (seals) them. Cooling is done in a continuous manner in cold flowing water in equipment similar to a continuous agitating sterilizer. If the cooling water is used repeatedly, it should be chlorinated to preclude infection of the can contents with spoilage organisms.

The cans are labeled and cased immediately after cooling and then stored the cans in the warehouse. The warehouse should be cool, dry; and well ventilated, and the cans must be protected against the accumulation of moisture to prevent rusting. In cold climates it may be necessary to heat the warehouse in order to prevent freezing of the cans or the condensation of moisture upon them with resultant rusting.

PIE Fruit

The trimmed, overripe, and otherwise low-quality fruit is canned as Pie fruit. Much of it is sliced before canning. It is heated in live steam for about 8 to 10 min., canned scalding hot in to cans, and sealed hot. If it is sealed at 195°F. or above, some canners find that additional processing is not necessary, if the hot cans are allowed to stand several minutes before they are cooled. Paneling of the cans handled in this manner is apt to be quite severe. Some canners prefer to give the sealed cans a process in an agitating continuous cooker for 30 to 40 min, to preclude any danger of spoilage.

Apricots (Prunus Armeniaca)

The Blenheim apricot is the most popular variety for canning purposes. It is of moderate size, of deep-yellow color, of excellent flavor, and reasonably free from scab or blemishes. When properly ripened, it has uniform texture from the skin to the pit and retains its shape in the can during processing.

The fruit should be harvested at its optimum degree of maturity for canning purposes. If it is too green the canned fruit will have a disagreeable astringent flavor and no amount of sugar will entirely overcome this defect. If it is overripe, it will be too soft and will be unattractive in appearance after sterilization. When gathered at the "canning-ripe" stage of maturity, the fruit is firm, of full size, of' good color, and of pleasing flavor. It will not have reached the maximum flavor, however, at this stage of ripeness.

The apricots are halved and pitted but are generally not peeled. The fruit is cut by hand around the pit suture, and the pits removed; or now more commonly a small machine fed by hand cuts the fruit. A small proportion of the crop is lye-peeled. Much fruit is now canned whole, some of it lye-peeled. Ripe fruit is graded for size, sorted, and canned without peeling or is lye peeled and canned. The whole fruit requires a longer period of processing than the halved, viz., 18 to 25 min., depending on size of the can and maturity. Considerable Pie-grade fruit is now pitted mechanically by Elliott Pitters, then steamed and canned as solid-pack Pie fruit, giving a superior product for this purpose. If it is canned boiling hot, additional processing is usually not necessary. Hand-pack filler usually cans it. In some canneries the fruit is canned from a broad, slowly moving belt.

The filled cans are syruped in the automatic syruping machines with syrups of the concentrations of 55, 40, 25, 10° Balling and plain water, according to whether-the grade is Fancy, Choice, Standard, Second, or Pie.

After exhausting and sealing, the cans are immediately sterilized, usually in continuous agitating cookers. The usual process is 10 to 15 min. Too prolonged sterilization will soften the fruit badly. Very little softening of the apricots by cooking is required, and for this reason the cans should be thoroughly and quickly cooled after sterilization.

Apples (Malus Domestica)

The canning of apples is considered a by-product industry in most apple growing districts and as a means of utilizing the best quality of culls. The fruit for canning purposes should be of fair size and reasonably free from blemishes. Apples unfit for canning may often be used for cider or vinegar. The two principal forms are quarters or segments in water and canned applesauce.

Apples for canning should be firm and hold their shape in the can. They should be of good flavor, color, and texture. Acid varieties of white flesh are preferred. On the Pacific Coast the Yellow Newtown Pippin, Winesap, Jonathan, and Spitzenberg are popular for canning purposes.

The fruit should be washed and sorted before it goes to the preparation tables. It is peeled by mechanical peelers, operated by power; the apples are placed on the peeling knives of the machine by hand. Ordinarily the fruit is put immediately in dilute brine to prevent oxidation and browning. The peels and cores, which normally represent from 30 to 40 per cent of the weight of the fresh fruit, are usually sent to the vinegar factory to be crushed and pressed for- vinegar, although some factories have found it profitable to dehydrate the peels and cores for the use of jelly and pectin manufacturers, from whom there was a good demand. Before the apples are canned they are usually blanched in one of several ways. A simple process of blanching consists in passing the quartered or sectioned apples through a steam box to soften them, to destroy the oxidase, and to expel the air from the fruit, thereby reducing spoilage. In some canneries the fruit is immersed in boiling 3 per cent brine for 3 or 4 min. It is possible to remove the air by placing the fruit in dilute brine or water and subjecting it to a high vacuum. The air is effectively removed, and water enters the fruit tissues to replace the expelled air, increasing the weight of the fruit considerably.

Following the blanching operation, the hot fruit is packed at once into cans as a solid pack, or a small amount of boiling hot water is added. In most cases, however, the can is practically filled in solid-pack style, and very little liquid is necessary. Type L or similar corrosion-resistant cans should be used. More trouble is experienced by the corrosion and pinholing of tin plate by apples than by other canned fruit.

Apples are easily sterilized on account of their high acidity, but because the fruit is packed tightly in the cans, heat penetration is not very rapid. Nevertheless, a sterilization of 8 to 10 min. at 212°F in a continuous agitating sterilizer has been considered sufficient if the cans have been filled and sealed above 160°F.

Blackberries (Rubus Villosus)

Moderate quantities of blackberries are canned for use in the preparation of pies. Freezing is supplanting canning.

Canning. At the cannery the fruit is generally merely sorted and very thoroughly washed, very little attempt being made to grade for size. Since most of the fruit is used for pie making rather than for dessert purposes, it is generally packed in water or in light syrups. Fruit for dessert purposes should be packed in syrup of 40 to 550 Balling.

Loganberries

Loganberries are used for canning, frozen pack, jams, and juice. The berries are very large in size and deep red in color.

The canned fruit is most in demand for pie-making purposes and therefore is canned. The processes of harvesting, canning, and sterilizing are practically the same as for blackberries. "Double-enameled" Type L cans should be used to ensure the retention of color.

Raspberries (Rubus Strigosus)

The red raspberry is preferred to the black variety for canning purposes but is more in demand for the preparation of preserves and jams than for canning. The berries are canned in lacquered cans, preferably of Type L plate, in heavy syrup for dessert purposes and in water for use in pies. The length of sterilization is usually about 12 min. at 212°F. They may also be canned after cooking a short time with half their - weight of sugar. They are usually frozen rather than canned.

Strawberries (Fragaria Virginiana)

Strawberries for canning purposes should be firm in texture, of good color and flavor, and of large size. The most important requirement is firm texture. The principal difficulty in the canning of strawberries is the softening of the fruit during sterilization, which results in the can containing only one-third to one-half its volume of berries. Strawberries are used for preserves and for frozen pack in very much larger quantities than for canning.

Strawberries shrivel if canned in too heavy a syrup, although a fairly heavy syrup is necessary to develop and retain the berry flavor. A syrup of 500 Balling is satisfactory. Strawberries are much more satisfactory for preserving and for frozen pack than for canning.

Cranberries

They are grown on land that can be flooded during the growing season, i.e., in cranberry bogs. They ripen in the fall and are picked by hand with a special rake-like device or by a machine that makes use of powerful air suction to strip the berries from the Vines and convey them to a cleaner and hopper. At the cannery the berries are cleaned by screening and winnowing to remove leaves, trash, etc.

Cherries (Prunus Cerasus)

This is a large, sweet variety of white or light pink color. After its arrival at the cannery in most plants, the fruit is first stemmed by hand or by mechanical stemmers. Most of the sweet cherries are canned without pitting, although most sour cherries are pitted. The pitting is accomplished by an automatic machine in which cherries fall into small cups and in which the seeds are removed by cross-shaped plungers.

Pears (Pyrus Communis)

The Bartlett pear is desirable because of its uniform shape, its white color, and its relatively small number of grit cells. Pears develop a better flavor and are of finer grained texture if ripened in boxes after picking. Fruit ripened on the tree is apt to be coarse in texture. The pears are gathered at full size, while they are still hard and green, and are shipped in this condition direct to the cannery where they are held for from 5 to 10 days to ripen. The fruit may be peeled, halved, and cored by hand, a special guarded knife being used for peeling.

Plums (Prunus Domesticus)

The large sweet varieties of white plums, such as the Green Gage and Yellow Egg, are in greatest demand for canning purposes; the red and black varieties are seldom used. The plums are sorted, and the stems and leaves removed, after which they are graded.

Oranges

The process used was to heat the sliced peeled oranges in a heavy syrup, approximately 50° Balling, for several minutes at 175 to 185°F. The thoroughly heated fruit and syrup were then placed in cans and pasteurized at not above 185°F.

The flavor of the oranges deteriorates after canning, and the product is not very palatable after several months storage.

Pineapples (Ananas Sativus)

Pineapple is not so common as fresh fruit, it is generally canned. The fruit is allowed to reach full maturity, not only for maximum flavor and quality but also so that the optimum canning cutout and yields may be attained. They then cut off the crown with a sharp knife and place the pineapple on a belt on an extended boom that carries it to a large bin on a special truck. The fruit bins from the fields are lifted mechanically from the trucks and conveyed to a central dumping station. The Ginaca machine cuts a cylinder from the center portion of each fruit, removes the shell, cuts off the shell portions at each end of the cylinder, and removes the core. An "eradicator" scrapes the edible flesh from the shell as completely as possible for use in crushed pineapple or for juice. Chain conveyer through a spray washer to the slicing machine carries the trimmed cylinders.

Canning of Vegetables

In general, we may consider canned vegetables as staple foods and the higher grades of canned fruits as dessert products. Vegetables differ from fruits in chemical composition, and for this reason the processes of canning differ from those used for fruits. Most vegetables contain more starch than sugar, as contrasted with fruits, which are high in sugar and low in starch.

The acidity of vegetables is generally much lower than that of fruits. Vegetables are grown in or near the ground as contrasted with fruits, which are grown generally on trees at a considerable distance from the soil. Vegetables, therefore, usually contain more of the resistant soil organisms than fruits and usually require more cooking than fruits to develop their most desirable flavor and texture. For these reasons vegetables in general require a more severe processing than fruits in order to render them sterile and to cook them sufficiently for table use. Vegetables often contain disagreeable flavors, or compounds, which should be removed before canning, or some must be wilted by heating before canning, and for these reasons they are usually blanched before being filled into the cans.

The demands of the canner have developed the varieties of vegetables most suitable for canning purposes. In general, vegetables for canning purposes should be uniform in color and quality, tender, and in prime condition. Because of the rapid deterioration of vegetables after gathering, they should be transported from the field to the cannery in as short a time as possible.

Artichoke (Cynara Scolymus)

The large flower buds resemble a green pine cone in appearance, and the fleshy bases of its scales, "leaves," and the receptacle, or hair, form the edible portion. At the cannery the artichokes are first graded for size by means of a pair of traveling, diverging endless ropes. The cans are filled in a prevacuumizing briner under vacuum with brine containing about 0.35 per cent citric add and 0.6 per cent salt. They are sealed in a vacuum-closing machine and processed in open 'retorts in boiling water' for 30 min and cooled to 100 to 140°F in water.

Green Beans (*Phaseolus Vulgaris*)

The beans should be of deep green color, crisp and tender, fleshy, and as free from fiber and strings as possible. The cut beans as well as the whole are blanched, usually in hot water at 185 to 190°F rather than in steam, for about 1 to 2 min. to wilt the beans slightly and thus give better-filled cans. At higher temperatures, or after a longer period at 180 to 190°F., the skins are apt to "sluff," giving a ragged appearance. As previously mentioned, small beans for cutting as French style are blanched before cutting. In this case the blanched whole beans are returned to the French-style cutters and cut into shoestring-shaped pieces. The cans are filled with hot water, and dry salt is added by dispenser; or a brine of 20 salometer is added to fill the can, and the can is heated in a steam exhaust box for about 1 min. The cans are sealed hot or are steam-flow-sealed.

Carrots (Daueus Earota)

Carrots are canned in diced, julienne, or shoestring style, sliced, halved, quartered, and whole. Varieties for canning should have a smooth contour free of deep folds or wrinkles, tender texture, and flesh of attractive orange color. The carrots should not have a white or light-yellow core, but the flesh should be well colored throughout.

Carrots store well in cool or cold storage if the humidity of the air is maintained at a high enough level to minimize evaporation of moisture. If the carrots "dry out" excessively, they become limp and rubbery and suffer in flavor. Canning soon after harvesting attains best quality.

Carrots and Peas - A mixture of diced carrots and green peas is canned in considerable quantity. They must be well mixed before canning and without undue crushing or breaking of the peas; consequently the peas should not be too heavily blanched before mixing. The ratio of carrots to peas is usually from 50:50 to 60:40.

Corn (Zea Mays)

The canning of corn is one of the most important of the vegetable canning industries, the average annual production being approximately 33 million cases.

OKRA (Hibiscus Esculentum)

Okra is used principally for soups and is a staple garden crop. The pods resemble peppers in appearance and are rich in gums and probably pectin. The pods should be picked while still tender and before they have become fibrous and tough. The pods may be blanched in water. The butts shou1d be cut off and discarded, and the pod should be cut into sections crosswise with a string bean cutter. The pods may also be cut in cross sections and canned without blanching.

Pumpkin and Squash

Canned pumpkin has made pumpkin pie available at all seasons of the year. Pumpkin for canning should be of the hard, sweet varieties, evenly ripened. The flesh should be golden yellow and of good texture, not watery. The Golden Delicious squash is popular for canning.

The pulp is separated from the skin and tough fiber by a heavy tomato pulper, and the pulp so obtained is usually passed through a finisher and then evaporated in open kettles to a heavy density and canned hot as a solid pack. Enamel-lined cans should be used to prevent darkening of the product. Exhaust is necessary if canned hot. The process is usually conducted at 250°F. for 75 to 90 min.

Rhubarb (Rheum Rhaponticum)

Rhubarb more clearly resembles a fruit than a vegetable in composition. On account of its high acidity it attacks tin plate vigorously, and perforation of cans is common. The rhubarb is washed and cut in short lengths. It is either packed at once into cans with water or is cooked a short time in a jelly kettle and packed hot into cans, the latter method being preferable because it gives a better-filled can.

Sweet Potatoes (*Ipomoea Batatas*)

The potatoes are washed and placed in crates in a retort at 240°F for 9 to 12 min. to soften and loosen the skins. A quick, high heating is better for the purpose than a longer cook at a lower temperature.

The steamed potatoes are taken at once to the peeling tables, where workers with heavy canvas gloves slip the skins from the potatoes. If properly steamed, the skins slip from the potatoes readily. The potatoes do not peel satisfactorily if they have been allowed to stand so long after digging that they have become shriveled. Revolving brushes are used in addition to Water sprays to remove the lye-softened skins. Hand trimming is usually necessary with lye-peeled potatoes. Lye-peeled potatoes or those peeled by machines are steamed to soften them before packing. The hot peeled potatoes are tightly packed at once into sanitary cans with the addition usually of a small amount of water. The cans are heaped full, and the potatoes are pressed in tight.

Tomatoes (Lycopersicum Esculentum)

Tomatoes are used in enormous quantities in the fresh state, as canned tomatoes, and in the form of juice, puree, paste, and various relishes, such as catsup, chili sauce, etc. Tomatoes at one time headed the list of canned fruits and vegetables in quantity. Tomatoes for canning should be moderately large, smooth, so that peeling can be easily accomplished, evenly ripened to the stems, of a clear red color, and possessing a large proportion of solid meat of good flavor. Those of irregular shape and wrinkled skins are difficult to peel, and the loss in preparation is excessive. Some varieties possess large seed cavities and on this account soften badly in the can giving an unattractive appearance and a slack-filled can. Soft, watery varieties are objectionable for similar reasons. Yellow and purple tomatoes are not desirable; a deep, uniform red color is the ideal.

Spoiling of Canned Foods

All canned foods, after sterilization, are subject to deterioration during storage. In many cases the changes that occur do not render the food unfit for consumption; nevertheless the appearance of the container or of the product may become so unattractive that it becomes unsalable. Two general types of deterioration are recognized: (1) spoiling by microorganisms, and (2) changes brought about by chemical or physical agencies. Microbiological spoiling of canned foods is the more important.

Spoiled cans of food exhibit characteristic differences in appearance, taste, and odor from normal cans. The most common designations in use for the different types of spoiled or abnormal cans of food are :

Swell. A swelled can is one whose ends are tightly bulged from the formation of gas within the can by microorganisms and the ends of which remain convex and spring back to this position if pressed inward. Swelled cans of food are usually so badly decomposed as to be unfit for consumption. They may be poisonous because of the presence of *Bacillus Botulinus* (*Clostridium botulinum*).

Hydrogen Swell. A hydrogen swell is caused by the formation in the can of hydrogen gas as a result of corrosion of the tin plate. The contents are usually sterile and often fit for food.

Springer. A springer may be merely a mild swell or a mild hydrogen swell. It may be caused by overfilling the can or by insufficient exhausting. Swells always pass through the springer stage. The ends, or at least one end, of a springer can be pressed in with the hand and will remain concave for a time. Springers, if caused by overfilling, underexhausting, or corrosion, can be used for food.

Flipper. A flipper is a can under very mild positive pressure. It may be of normal appearance, but the end, if struck sharply against the top of a table other solid object, will become convex. It may represent the initial stage a swell or hydrogen swell or, more frequently, be due to overfilling or under exhausting.

Flat Sour. A flat sour is a can of food which has undergone spoilage by microorganisms without gas formation and is normal in outward appearance.

The product usually possesses a sour taste and frequently a sour odor. Vegetables frequently undergo this type of decomposition, which is usually a nonpoisonous change. Occasionally this condition is caused by bacterial spoilage before canning.

Leakers. Cans may leak because of (*i*) faulty seaming of the factory end of the can, (*ii*) faulty seaming of the cannery end of the can, (*iii*) faulty lock seaming, (*iv*) pinholing by corrosion from the inside of the can or rusting of the outside of the can, (*v*) bursting of the can by "excessive gas pressure developed in the can by decomposition, by microorganisms, or by formation of hydrogen gas through corrosion, and (*vi*) by rough handling of sealed cans.

Unfermented Fruit Beverages

The production of fruit juices has increased greatly in recent years. Large amounts of fruit concentrate and syrup are used as a base for carbonated bottled beverages and for a bottled noncarbonated beverage; in both cases the syrup or concentrate is diluted greatly with water before bottling, and the products are preserved with sodium benzoate. During the Second World War a great deal of orange concentrate was exported to Great Britain to be used by children as a source of vitamin C. Very large amounts of citrus and other fruit concentrates are preserved by freezing. Most of the preserved juices on the market are cloudy or pulpy or clarified with a pectic enzyme. The fruit should be picked at the proper stage of maturity for the preparation of juice, which will vary with the variety. Thus loganberries should be picked when they have become "dead" ripe, i.e., soft ripe, for they are then at their optimum color and flavor.

Preservation of Fruit Juices

Several methods are in used for the preservation of fruit juices. The most important of these are discussed below.

Pasteurization. Pasteurization as applied to fruit juices means the destruction, by heat, of all microorganisms capable of increasing in the juice and of causing spoiling. It usually does not kill the spore bearing organisms, such as thermophiles, *Bacillus Subtilis*, *B. Mesentericus*, etc., but these organisms and most other spore-bearing bacteria as well cannot grow in acid fruit juices, and consequently their presence is of no practical significance. Pasteurization of still (noncarbonated) juices need only be at—such a temperature and for such a time that yeasts and molds are destroyed. Yeast is killed by heating for a few minutes at 140 to 180°F.

Bulk Pasteurization. It is often necessary to store fruit juices in bulk in large glass carboys or in barrels to permit settling or shipment in bulk. Two types of pasteurizers, which may be designated as (1) continuos and (2) discontinuous, are used for this purpose.

Grape Juice

Clear fruit juices are very attractive in glass containers, a fact that aids naturally in direct advertising to the consumer.

Apple Juice

Unfermented apple juice (sweet cider) is one of the most popular fruit juices. Much of this juice is consumed fresh, directly after pressing, or from barrels in which it is preserved with benzoate of soda. Benzoated cider is often of very poor quality, and usually the flavor of benzoate is evident. The sale of juice preserved in this manner is, in the author's estimation, the principal obstacle to expansion of the sweet cider industry.

Apple juice should possess a rich apple flavor and should be tart. Although apple juice is sometimes considered a by-product and a means of utilizing culls, the raw material should be sound, free from rot, worms, and fermentation. Apples to be used for juice should in all cases be thoroughly washed before crushing, because even under the best conditions they will carry considerable dust and are frequently contaminated with juice or pulp from spoiled fruit. The apples may be soaked by conveying them through a long tank of running water, and the loosened dirt may be effectively removed by sprays.

However, juice producers find that it is advisable to determine by small scale tests the best concentration of enzyme for their conditions of temperature and variety of apple.

Packaging and Pasteurization. Until fairly recently most fruit juice, including apple, was pasteurized in the bottle or can after filling cold and sealing. Large shallow wooden vats equipped with slat false bottoms were used. The bottles were placed, preferably on their sides, in the vat, and water was added to cover and heated to 105 to 180°F for a long enough time to prevent subsequent spoilage by mold.

Carbonating. Apple juice is greatly improved for the average consumer by carbonating, and some of the bottled juice now on the market is lightly carbonated, i.e., at 15 to 20 lb. pressure per sq. in. During the Prohibition era some breweries were converted into juice factories in some apple-producing localities, particularly in the Pacific Northwest, and the carbonating, bottling, and pasteurizing equipment formerly used in these establishments for beer was used successfully for cider.

The Schwartz Process. The well-known Sand W Co. of Redwood City, California, bottles and cans a cloudy apple juice of remarkably fresh flat and light (almost white) color. The washed and sorted apples are juiced in a Schwartz hammer mill-type juice extractor under vacuum. The juice is further vacuumized, flash-pasteurized, canned or bottled hot, and cooled.

Juice Extraction. Various methods of extracting the juice can be used, but the following is recommended. The berries are crushed in a stainless steel crusher, heated, and pressed in a rack-and-cloth cider press. In most cases the berries are heated to 160°F. before pressing to increase the yield and to obtain a juice of more intense color. Juice from unheated fruit contains less pectin and is therefore more easily filtered,

Filtration. Filter presses or pulp filters may be used for filtering the juice. Because of its high content of gums and pectin, it is difficult to filter. Addition of infusorial filter aid is advisable. It is not necessary to place the juice in cold storage to permit settling before filtration. It is feasible to filter the juice satisfactorily after 24 hr. settling at room temperature following preliminary pasteurization at 160 to 165°F. to coagulate proteins. Clarification with a pectic enzyme greatly aids filtration.

Citrus Juices

Until about twenty years ago all attempts to pack commercially a satisfactory canned or bottled orange or lemon juice were unsuccessful because of undesirable changes in flavor and odour during storage.

Frozen Orange Juice. Some orange juice is preserved by freezing in Darnel-lined cans for use by ice-cream producers, hospitals, steamships, and other large users of fruit juices. Attempts to distribute frozen orange juice in the retail trade or direct to the home have not been very successful, although the product is satisfactory. Apparently house W lives prefer to use the fresh oranges or the canned juice, as the cost for an equivalent amount of ice is considerably less than that of the juice distributed in the frozen state. Some frozen orange juice in special paperboard cartons was once marketed.

Apricot Nectar. This product has become rather popular. Its production was initiated as a result of experiments made by A. Shallah and others at the University of California. The following procedure was used: The apricots (thoroughly ripe) were pitted, steamed until soft, sieved tomato-juice machine, and mixed with an equal volume of 15° Brix syrup. The mixture was heated to 180 to 185°F., canned hot sealed, processed at 212°F. for 20 min. in cans, and cooled. This procedure retains vitamin C, whereas cold sieving of the fresh raw fruit may lose it.

Prune Juice. This is a water extract (infusion) of dried prunes. Leonard Lane, Ponting, and the author found the following procedure to give an excellent juice. Dehydrated prunes of best quality were heated to 175 to 180°F. with about four times their weight of water for several hours. The resulting "juice" was placed on a second lot of prunes, and heating as repeated. Three such operations were made.

Methods of Preservation of Fruit Juices

Several methods are in commercial use for the preservation of fruit juices. The most important of these are as follows.

Pasteurization. Pasteurization as applied to fruit juices means the destruction, by heat, of all microorganisms capable of increasing in the juice and of causing spoiling. It usually does not kill the spore bearing organisms, such as thermophiles, *Bacillus Subtilis, B. Mesentericus,etc.,* but these organisms and most other spore-bearing bacteria as well cannot grow in acid fruit juices, and consequently their presence is of no practical significance. Pasteurization of still (noncarbonated) juices need only be at—such a temperature and for such a time that yeasts and molds are destroyed. Yeast is killed by heating for a few minutes at 140 to 150°F., and resistant mold spores will require in most cases a temperature of 175°F. for 20 min. Molds require-oxygen for growth, and for this reason heavily carbonated juices can be pasteurized. Safely at 150°F., which destroys yeast cells. Most still juices must be pasteurized at 175°F juices of high acidity may be pasteurized at a lower temperature, 160 to165°F.

Clarification of Fruit Juices by Settling, by Fining, and by Enzymes

It is possible in certain cases to clarify fruit juices by settling, with or without pasteurization as the juice may require, or by the addition of fining materials.

Settling. Frequently fruit juices after pasteurization will become clear during storage, the length of storage necessary depending upon the variety of juice and other conditions. Thus pomegranate juice will become clear within 24 hr. after pasteurization, while grape juice usually requires several months settling. In the commercial manufacture of grape juice, settling of the pasteurized juice greatly facilitates filtration by eliminating much suspended matter.

Use of Finings. Some juices that do not settle satisfactorily during storage and are difficult to filter can be clarified by the addition of 3 fining agent. This agent may be defined as a substance that, when added to the liquid to be clarified, will form a precipitate which settles and carries with it the finely divided particles responsible for the cloudy appearance. The fining materials most commonly used for fruit juices are egg albumen, casein, bentonite clay, and infusorial earth.

Pectin, Jellies, and Marmalades

The manufacture of jellies and marmalades is one of the oldest and most important of the fruit-products industries and affords a means of utilizing a large amount of sound fruit unsuited to other purposes.

Jelly. Jelly is prepared by boiling fruit with or without water, expressing and straining the juice, adding sugar (sucrose), and concentrating to such consistency that gelatinization takes place on cooling. A perfect jelly is clear, sparkling, transparent, and of attractive color. When removed from the glass, it should retain its form and should quiver, not flow. It should not be syrupy, sticky, or gummy and should retain the flavor and aroma of the original fruit. When cut it should be tender and yet so firm that a sharp edge and smooth, sparkling cut surface remain.

Marmalade. A true fruit marmalade is a clear jelly in which are suspended slices of fruit or peel. Frequently jams are mislabeled as marmalades.

Constituents of Jelly

Three substances are essential to the preparation of a normal fruit jelly. These are pectin, acid, and sugar. Of these, pectin is the most important.

Pectin. It is possible to make a jelly of excellent consistency by combining pectin, acid, sugar, and distilled water in the proper proportions. Fruit juices that are normally deficient in pectin or acid, or both, will make good jelly if these constituents are added.

Acid. Acid is a necessary constituent of fruit jellies. Juices that are deficient in acidity will make good jelly if citric, tartaric, or other suitable acid is added, provided the proper proportions of pectin and sugar present.

Sugar. Sugar, the third necessary constituent of fruit jellies, may be in the form of any readily soluble sugar, such as cane sugar, dextrose, levulose, maltose etc. Jelly forms when the concentration of the water-sugar-acid pectin mixture attains a certain minimum value, which is dependent within limits on the proportions of pectin, acid and sugar.

Nature of Jelly

It is probable that the formation of jelly from pectin, acid, and sugar is not a definite stable chemical compound, because if fruit jelly is diluted with warm water the constituents go into solution and can be separated by suitable physical means.

Suitability of Various Fruits for Jelly

The jelly should contain sufficient acid and pectin to yield a good and stable produce. Some fruits contain enough of both pectin and acid for the purpose; some are deficient in one or the other; and some are deficient in both substances.

Of the fruits rich in pectin and acid, crab apples, acid varieties of table apples, loganberries, sour blackberries, currants, lemons, limes, grapefruits, sour varieties of oranges, sour varieties of guavas, damson plums, most other varieties of sour plums, Labrusca varieties of grapes, sour varieties of cherries, cranberries, and roselle are good examples. Of fruits and vegetables low in acid but rich in pectin, the following may be cited: sweet varieties of cherries, unripe figs; ripe melon, carrots, unripe bananas, and ripe quinces. Fruits and vegetables that are rich in acid but low in pectin are apricots, rhubarb, and most varieties of strawberries.

Fruits that may be classed as containing a moderate concentration of both acid and pectin are ripe Vinifera varieties of grapes, ripe blackberries, ripe apples, loquats, and feijoas.

Fruits low in both acid and pectin are represented by pomegranates (arils), ripe peaches, ripe figs, and ripe Bartlett pears.

It is customary to blend fruits deficient in acid or pectin, or both, with fruits that have an abundance of the required constituents. Because it possesses no appreciable flavor of its own, commercially prepared pectin is coming into more general use for the purpose of enriching juices of fruits deficient in this component.

The Preparation of Jelly

The jelly is prepared by several steps, viz., those of boiling the fruit, extraction of the juice, clearing the juice, adding the sugar, boiling, packaging, and sterilizing.

Boiling. Most fruits should be boiled for extraction of the juice in order to obtain the maximum yield of juice and pectin, because boiling converts pectose into pectin and softens the fruit tissue. Very juicy fruits, such as berries, do not require the addition of water

Water. The amount of water that should be added to the fruit should be sufficient only to obtain a good yield of juice and pectin. Juicy fruits require no water; apples require from one-half to an equal volume of water; and citrus fruits, because of the long period of boiling necessary, usually require from two to three volumes of water for each volume of sliced or crushed fruit. If too much water is used, the resulting juice will be too dilute and will require an undue amount of concentrating before jelly can be made from it; if too little water is used, there is danger of scorching the fruit or of obtaining a low yield of juice and jelly. Fruits very rich in pectin, such as currants, loganberries, cranberries, lemons, and Labrus varieties of grapes, can be extracted to advantage with two or more succeeding lots of water. Frozen-pack berries and plums used for jelly making are handled in much the same manner as the fresh fruits. They should be dumped into the required amount of boiling water while the fruit is still in the frozen condition.

Kettles. The extraction of juice from fruits for the preparation of jelly in a commercial scale is usually accomplished by the use of steam-jacketed kettles, placed on a platform or a floor above the press so that the cooked pulp and juice may be drawn from the kettle by gravity to the press. If large kettle (50-gal. capacity or more) is used, it should be fitted with a large valve (2 in. or larger) to permit drying off of the fruit.

Pressing. The housewife, in preparing juice for jelly making, usually does not press the fruit; she merely places the heated pulp and juice in a cloth jelly bag and allows it to drain, in order to obtain a clear juice.

In the jelly factory a high yield of jelly juice rich in pectin and obtained with a minimum of handling is desired. The use of the rack-and-cloth press has been found in practice to be one of the most desirable means of pressing the juice from the boiled pulp.

Use of Pomace. The press cake may, if desired, be mixed with water in the kettle and heated a second time to obtain the remaining pectin. This is probably not advisable for cheap fruits, such as apple culls and apple waste or citrus-fruit culls, but may become profitable with more costly fruits, such as currants, loganberries, etc.

The press cake (pomace) has some value as stock food and can be fed direct, or it can be dried, stored, and used as needed. It has very little fertilizing value but can be used to improve the texture of heavy soils if mixed with lime in order to prevent formation of harmful concentrations of acid in the soil.

Clearing the Juice. Jelly is most attractive when clear, and most jelly factories not use mechanical filters.

Filtration. Filtration must be accomplished before the addition of sugar, because the latter so increases the viscosity of the juice that filtration becomes extremely slow or impossible. If the juice requires concentration by boiling before the addition of sugar, this should be done before filtering, since boiling causes precipitation of organic matter (probably protein), which should be removed by filtration or other means before sugar is added.

Settling. Some fruit juices can be satisfactorily cleared by settling overnight in vessels 1 to 3 ft. in depth. Shallow tanks should be used because of the relatively slow rate of settling of juices from boiled fruits.

Centrifuging. Experiments with centrifugal clarifiers have proved that jelly juices can be partially clarified by centrifuging at a high speed. As much of the coarse pulp as possible should be removed before centrifugal clarification, in order to avoid too rapid clogging of the bowl of the clarifier. This method is rapid and inexpensive in operation. The centrifuged juice should be filtered if very clear jelly is desired.

Cooling. Some continuous pasteurizers are equipped with a device for cooling the jars after pasteurization. This may consist of several tanks of progressively decreasing temperature, through which the jars pass, or of a series of sprays of water of progressively decreasing temperature. If allowed to stand hot for several hours, the jelly will be materially injured in flavor and color.

The most common procedure consists in filling the containers at about 190 to 200°F. The jars or glasses are usually preheated. After filling and capping, the containers are sprayed with hot water to remove jelly adhering to the outside and lids before it has solidified. They are then usually can put through sprays of water of progressively decreasing temperature to cool rapidly to room temperature or slightly above.

Deciduous-fruit Marmalades. Marmalades are also made from other fruits, although many so-called "marmalades" are jams rather than marmalade's. Various sliced fruits can be mixed with a juice rich in pectin and sugar in preparing a true marmalade, i.e., a jelly in which are suspended pieces of fruit. The famous Bar-Ie-due of France is essentially a marmalade prepared from currants.

Preparing the Juice for Marmalade. According to the usual American factory practice in making marmalade, the juice and the sliced fruit are prepared separately and are not mixed until the final boiling of the juice and fruit with sugar. .

Slicing. In preparing marmalade from oranges and lemons, these fruits are mixed in the proportion of about 1 to 5 of lemons to 4 to 10 lb. of oranges and sliced about 16 in. thick. Ripe fruits of both varieties are used.

Boiling. The sliced fruit is covered with 2 to 3 times its volume of water in a jelly kettle (glass-lined or stainless-steel equipment is to be preferred). The mixture is boiled until the fruit is tender, usually about 1 hr. It is sometimes necessary to add water during boiling to replace that lost by evaporation.

The hot pulp is then pressed in a rack-and-cloth type of press. Heavy cloths or two thicknesses of ordinary press cloths should be used in order to eliminate as much of the fine fruit pulp as possible.

Filtration. Settling in shallow vessels for 24 hr. can clear the juice by filtration. Felt or heavy duck bag filters yield a juice which is opalescent but which, nevertheless, produces a marmalade of satisfactory appearance. By use of a filter press it may also be filtered hot after addition of infusorial earth.

Preparation of Marmalades

A good marmalade should be a jelly with pieces of fruit suspend therein and should not be merely a jam or butter. The principles of jelly making, therefore, apply also to the preparation of marmalade.

Types of Marmalade. English and Scotch marmalades are usually made from the bitter varieties of oranges. In America sweet varieties are used.

Fruit for English Marmalade. The fruit from which the English type of marmalade is produced is high in both acid and pectin, and no difficulty is experienced in obtaining a firm, jelly like marmalade from it.

Fruit for American Marmalade. In the United States, marmalade is usually made from cull oranges. The product is characterized as sweet marmalade distinguished from the bitter English marmalades.

Preparing the Sliced Fruit. For the preparation of the usual English marmalade the whole fruit is used, and the juice and peel are not prepared separately. The fruit is very finely shredded by a special machine designed for this purpose.

Three methods of preparing the peel are in use in marmalade factories in California. In one method a band of peeling about 1 in. wide is cut from the orange around its greatest circumference. This band is then cut crossing into very thin slices about 3-2 in. thick. The pieces possess a "shoe-peg" appearance and give a very attractive marmalade. In one large English factory the peel is removed in three or four segments, which are then cut in thin strips. In another factory the whole fruit is sliced.

In another method some of the whole fruit is sliced very thin and boiled until tender. It is then placed on screens, and the yellow rag and pulp are washed from the peels.

In one large factory the hole fruit is chopped finely by means of a mechanically driven mincemeat chopping bowl. As in the English process an attempt is made in this case to prepare the juice and peel separately. Marmalade prepared according to this method is cloudy and of jam like rather than jellylike consistency, but its flavor is excellent.

Boiling and Packing. The juice and peel are combined after the later has been boiled in water until tender. The proportion of peel to juice will depend upon the pectin content of the juice and upon the thickness of the peels. Where the slices are dry thin and the juice is rich in pectin, about 5 to 7 per cent of the sliced peels may be added to the juice, together with sugar equal in weight to the juice. If the slices are relatively thick, a larger proportion by weight of peel can be added.

Where the whole chopped or sliced fruit is used without previous separation of the peel and juice, the fruit should be boiled until tender before sugar is added.

Addition of Sugar. The amount of sugar that is required varies greatly with the composition of the juice. As in jelly making, a relatively greater proportion of sugar can be added to juices rich in pectin and acid than to those deficient in one or both of these constituents. Equal weights of juice (i.e., juice and fruit) and sugar is the normal proportion.

Use of Pectin. By addition of the required amount of pectin, results will be more uniform and there will be less danger of the failure of the marmalade to gel. Another great advantage is that the formula can be standardized in respect to ratio of sugar to the other ingredients.

Quick-setting pectin is preferable to the slow setting in this case as it will gel rapidly enough to prevent undue separation of the peel and jelly.

End Point. The juice, peel, and sugar, or sugar and sliced or chopped whole fruit, are boiled to the jelling point, usually 219 to 220°F., or to the desired total solids content as determined by refractometer. The tests previously described for determining the finishing point of jellies can be used in the case of marmalades. A good marmalade should not be syrupy but should be of jellylike consistency.

Cooling. Marmalade should be allowed to cool partially and to stand a short time to permit absorption of sugar by the peel from the surrounding syrup before the marmalade is placed in the final containers, unless the whole fruit is used without previous separation of juice and peel. If the marmalade is packed boiling hot direct from the jelly kettles, the peels are apt to come to the surface instead of remaining in suspension. However, with use of quick-set pectin separation is not apt to occur.

Flavoring. The boiling of marmalade removes a great deal of the orange oil from the peels, and the finished product, if made from commercial sweet varieties of oranges, is liable to be lacking in distinctive flavor. A small amount of orange oil or orange extract added to the marmalade and mixed with it thoroughly after the boiling has been completed will usually considerably improve the flavor.

Pasteurizing. The marmalade should be sealed in glass or tin at about 150 to 180°F., as described elsewhere for jellies. Vacuum-sealed containers are best for the purpose, because they reduce the tendency of the product to oxidize. They should be pasteurized in water at 180°F. If filled and sealed at or above 185°F., pasteurization is not necessary.

Other Marmalades. Excellent marmalade can be prepared by combining apple juice rich in pectin and acid with thinly sliced firm peaches, with figs similarly prepared, or with other firm fruits. The juice and sliced fruit can be mixed with sugar and concentrated to the jelling point in the usual manner.

Causes for Failure in Jelly Making

Too Much Sugar. The usual cause for failure is the addition to the juice of too much sugar in proportion to the pectin and acid of the juice. Firm jelly can be obtained by properly adjusting the proportion of sugar to the pectin and acid.

Prolonged Boiling. Too prolonged boiling results in the hydrolysis of the pectin and in the formation of a syrupy caramelized mass. The juice and sugar should be concentrated to the jellying point as rapidly as possible in order to avoid hydrolysis of the pectin.

Crystals. At ordinary temperatures jelly may develop sugar crystals if the concentration of the finished product exceeds 70° Balling. During the normal boiling of jelly, some of the cane sugar is hydrolyzed to dextrose and levulose, which exhibit less tendency than cane sugar to crystallize.

Use of Dried Fruits in Preparation of Jellies and Marmalades

Large quantities of dried apple peels and cores from evaporating plants and of dried pomace from apple cider and vinegar factories have been used in the commercial manufacture of cheap jellies for bakers use. The apple juice is prepared by soaking the dried material overnight, followed by boiling, pressing, and filtering as for the preparation of juice from the fresh fruit. It is then usually combined with berry juice red-grape juice, or artificial flavor and color are added.

Fruit Jams, Butters, Preserves, and Confections

Fruit jams, butters, and preserves are very generally used throughout the civilized world. The people of the British Empire are probably the most important manufacturers and consumers of jams; France is noted for preserved and candied fruits of high quality; China produces preserved ginger root and candied fruits, which are well-known articles of world commerce; and India is famous for a fruit relish known as "mango chutney."

Housewives prige themselves upon their preserves and jams, and probably a greater quantity of fruit is used for these purposes in the home than is used in the factory production of jams and preserves.

Jam. Jam is prepared by boiling the "whole fruit pulp with sugar (sucrose) to a moderately thick consistency without retaining the shape of the fruit. In England a jam is usually considered to consist of fruit pulp cooked with sugar to a jelly consistency.

Fruit Butter. This product is prepared by boiling the screened fruit pulp with or without the addition of sugar, fruit juices, and spices to a semisolid stage of homogeneous consistency. It differs from jam in being of higher concentration and finer consistency. It is usually heavily spiced and is frequently prepared without the addition of sugar.

Fruit Paste. Fruit paste or fruit leather, etc., is prepared as described for fruit butter, but is dried in the sun or by artificial heat to a solid consistency or to approximately the consistency of putty.

Fruit Preserves. Preserves are made by cooking the prepared fruit in sugar (sucrose) syrup until the concentration of sugar reaches 55 to 70 per cent. A limited proportion of the sugar used may be in the form of Corn syrup. The fruit should retain its form, should be crisp rather than soft, and should be permeated with the syrup without shriveling of the individual pieces. Government regulations require the use of not less than 45 lb. of fruit for each 55 lb. of sugar.

Candied Fruits. Candied fruits are prepared by gradually concentrating fruits in syrup by repeated boiling until the fruit is heavily impregnated with sugar, this process being followed by drying to overcome stickiness.

Glace fruit is prepared by coating candied fruit with a concentrated solution of sugar and confectioners' glucose syrup, followed by careful drying to give a transparent glaze to the surface.

Fruit Confections. This is a general term used to describe candies in which fruits are used. There are on the market a large number of products of this character which vary greatly in appearance, texture, flavor, and in the proportion of fruit used in their manufacture.

Jams

Jams may be made from practically all varieties of fruits and from some vegetables, In the United States and in the British Empire the small fruits and berries are most popular for the purpose. Various combinations of different varieties of fruits can often be made to advantage, pineapple being one of the best for blending purposes because of its pronounced flavor and acidity. Fruit for jam making should have reached full maturity in order to possess a rich flavor and be of the best desirable texture.

All berries must be carefully sorted and washed; strawberries must be stemmed; peaches, pears, apples, and other fruits with heavy skins must be peeled; while apricots, plums, and other thin-skinned fruits do not require peeling, apricots, plums, and fresh prunes can be pitted by chine, such as the Elliott pitter. Firm fruits should be boiled in a small quantity of water before sugars added in order to facilitate pulping.

Frozen-pack fruits are melted quickly in the jelly kettle. Those preserved in sulfur dioxide solution are boiled with some water until most of the sulfur dioxide is volatilized, Berries should not be softened by boiling before the addition of sugar. In some plants it is customary to pass the cooked benoies that possess prominent seeds through a pulpier to give a seedless puree.

Addition of Sugar. Pure fruit jam as defined by the old pure food and drug regulations contains only fruit and cane or beet sugar (sucrose). Glucose in limited amount may also be used under the present regulations. The proportion of sugar to fruit varies with the variety of the fruit, its ripeness, and the effect desired. This is usually a suitable ratio for berries, currants, plums, apricots, pineapple, and other tart fruits, Sweet fruits of low acidity, such as ripe, peaches, sweet prunes, and vinifera varieties of grapes, normally require less than an equal weight of sugar, and the ratio may in some cases be as low as lb. of sugar per pound of fruit.

Boiling. Boiling is desirable in order to cause intimate mixing of the fruit pulp and the sugar and to partially concentrate the product by evaporation of excess moisture.

Berries should be used in small lots and concentrated to the desired consistency as rapidly as possible. Other fruits are more resistant to the action of heat and may be boiled more slowly or in larger lots.

End Point. Lost jams should be concentrated to a boiling point of 219 to 221°F., the end point varying with the fruit variety, proportion of sugar, and other factors a jelly thermometer may be used with advantage to determine the end point of the boiling process. An Abbe refractometer is very useful in determining the finishing point. The usual finishing point is 60 to 68 per cent soluble solids as determined by Abbe refractometer,

Use of Pectin. The combining of fruit pulp with pectin is becoming a more general practice in the commercial manufacture of fruit Jam in order to obtain products of jelly like consistency.

Fruit Butters

The most important fruit butter produced is apple butter; peaches and plums are used to a limited extent for the same purpose. In continental Europe plums and prunes are used in very large quantities for the preparation of cheap butters and highly concentrated jams. In the commercial manufacture of apple and peal butters, the fruit is cooked until soft with a small amount of water and without previous peeling or coring. Toxic spray residues, if present, must be removed, or the fruit peeled, before cooking and sieving. The softened fruit is then passed through a tomato pulper to remove skins and seeds and generally through a tomato-pulp finisher to impart a smooth texture.

End Point. The finishing point is determined by consistency rather than by boiling point, because the boiling point of the finished butter is dependent upon the ratio of pulp to juice and of pulp to sugar. The butter, when cold, should be thick enough to stand, then a spoonful is placed on a plate and should not flow. It should, however, be thin enough to spread easily on bread.

Fruit Butters with Sugar. In the preparation of fruit butters from pears and peaches, sugar is generally used instead of fruit juice. Brown sugar is often substituted for refined sugar because a finished product of dark color is usually desired. The usual proportion is % lb. of sugar per pound of pulp. The mixture is then concentrated to a heavy consistency and spiced as described above.

Canned Sauces

Apples are peeled, cored, quartered, steamed, pulped, and cooked, with 1 part of sugar to about 6 of fruit. The sauce is usually sieved before canning. It is canned hot, sealed, and processed for a short time at 212°F. If it is canned and sealed at 185°F. or above, further processing is unnecessary. Other fruit sauces may be prepared in similar fashion.

Preserves

Fruit preserves of good quality should retain the form of the original fruit and should consist of the whole or cut fruit in a clear syrup of high sugar concentration. The fruit should not be caramelized by overcooking and should retain most of the color of the fresh fruit.

Preparation of the Fruit. Fruit is prepared for preserving as for canning. Some fruits, as peaches and pears, are also sliced.

Open-kettle One-period Process. The usual process for the preparation of fruit preserves consists in boiling the fruit in steam-jacketed kettles with sugar or in syrup until a syrup of heavy density is obtained and the fruit is thoroughly impregnated with the syrup.

The syrup should have a final concentration for most fruits of 65 to 68° Balling, or the boiling point should be approximately 219 to *221°F.* at sea level.

The objection to the open-kettle one-period process is that it may result in serious injury to the flavor and color of the finished product. Its advantage lies in its rapidity and low cost of operation.

The Slow Open-kettle Process. In order to avoid undue injury to the color and flavor of the fruit the slow process of preserving may be employed. In this the fruit is heated for a short time on successive days in a syrups of progressively increasing sugar concentration.

The fruit, is heated to boiling in this syrup and allowed to stand overnight. The syrup is then increased to 60° Balling by the addition of sugar. The fruit and syrup are heated for 3 or 4 min. and are packed boiling hot into jars or cans. Normally no further processing is necessary, although if the syrup and fruit cool to below 180°F. before packing in the final containers, the product should be pasteurized at 180°F. for 30 min. or processed at 212°F. for 10 min. to ensure against spoilage. If the spices are tied in a cheesecloth bag, they may be removed from the syrup after preserving is completed. Spice oils and extracts are much more convenient than the dry spices.

Candied Fruits

The manufacture of candied and *glace* fruits has in the past been a highly specialized industry in which success has depended to a very great extent upon the skill and experience of the individual workmen. A very large proportion of the work has been done by hand labor and with small individual lots of fruit. Undoubtedly, the processes for the candying of fruits lend themselves to factory methods, and existing factory equipment could be employed to greatly reduce the cost of manufacture and the price to the consumer.

General Principles. The candying process consists essentially of slowly impregnating the fruit with syrup until the sugar concentration in the fruit is high enough to prevent spoiling. The process must be so conducted that the fruit does not soften and become jam or become tough and shriveled. Repeated boiling and storage in syrups of progressively increasing sugar concentration will accomplish the desired results. Following impregnation, the fruit is washed and dried. It may be packed and sold in this form, or it may be coated with a thin glaze of sugar and glucose *syrup-glanced.* This is done by dipping the dried candied fruit in a syrup and again drying the surface.

Preparing Fruit. Frequently the fresh fruit is stored in a dilute solution of sulfurous acid, or sulfur dioxide and lime solution, in order to bleach the color, harden the tissues, and to preserve it until needed, as described for Maraschino cherries.

Whole Fruits. Fruit preserved in sulfurous acid must be thoroughly bleached repeatedly in hot water to remove all taste of sulfur dioxide before the candying process is begun. Cherries are stemmed and carefully pitted before leaching. Apricots should be pitted without cutting the fruit in half. Small pears, plums, prunes, and other whole fruits are often pricked with copper wires.

Fresh fruits and canned fruits can be used for candying purposes without the intermediate step of storage in sulfurous acid. Figs, peaches, pears, and pineapple are particularly well suited for use fresh.

Use of Canned Fruits. All firm varieties of canned fruits may be used very satisfactorily. Pine apple, peaches, figs, and pears are particularly desirable for the purpose. Frequently fruit which is at the best stage of maturity for table use is too soft for the preparation of candied fruits.

The jelly was also cast in confectioner's starch molds in the same manner as ordinary gumdrops. On standing 24 hr., the cast pieces were separated from the starch and coated with chocolate or with coarse sugar. In the latter case the pieces were moistened with a wet towel and rolled in coarse confectioner's sugar.

A typical formula used in semi commercial tests is as follows:

	Pounds
Fruit pulp	50
Cane sugar	45
Invert siMIP	45
Powdered pectin, 150 grade	1.5
Citric acid	0.5

Dissolve the pectin by mixing it with 10 lb. of the cane sugar and adding it to about 2 gal. of water at 130 to 140°F., stirring, and finally boiling a short time. Dissolve the acid in 1 pt. of hot water. Add the pectin solution to the fruit in a jelly kettle. Mix well. Add the remaining sugar and invert syrup. Boil to 222°F., that is, to boiling point. Add the citric acid solution at this point. Stir well. Cook to 223°F and cast in starch or pour on an oiled slab to harden.

Dehydration of Fruits

Just as the Civil War stimulated the canning industry, the Boer War and the First world war stimulated the dehydration industry. To conserve cargo space and transportation facilities, enormous quantities of foods were dehydrated during both world wars and shipped to the Allied armies in Europe. In Germany in 1898 there were, according to Prescott, only, three drying plants. In 1917 the number had increased to about 1900, a fact that explains in part Germany's ability to maintain her food supply during the war. Similar expansion occurred from 1939 to 1944.

The dehydration of fruits has become a well-established and growing industry. Apples have been dehydrated (or evaporated) in America for at least a century, and the dehydration of prunes in the Pacific Northwest is an old and formerly very important industry.

Dehydration is at present defined industrially as drying by artificially produced heat under carefully controlled conditions of temperature, humidity, and air flow. To dehydrate means to remove water.

The term "dried" is applied to all dried products, regardless of the method of drying.

Evaporation is usually considered industrially to mean drying under conditions of humidity, temperature, and air flow not so carefully controlled as in dehydration. Evaporation is a broader term than dehydration. If applied literally, the evaporation of fruit would mean the complete vaporization of the whole fruit, both water and solids. Alcohol, ether, and other liquids, as well as water, may be evaporated.

Desiccation is essentially equivalent in meaning to dehydration. The terms "drier," "dehydrater," and "evaporator" are at present used more or less indiscriminately, although the term "drier" is often considered to be a general one applicable to all types of drying apparatus and a dehydrator is often considered to be more efficient and to permit of more exact regulation than an evaporator.

Pickles

The manufacture of pickles, relishes, and condiments is one of the most important of the food industries. Although the preservation of vegetables and fruits in pickled form began as a household art, at present most of the world's supply of pickles is produced in commercial plants.

Types of Pickles. The cucumber, one of the most important raw materials used for pickles, is packed in many forms, e.g., in plain or spiced sweet vinegar in jars, kegs, or cans or fermented in spiced brine as dill pickles; packed in mustard; or in chopped form in various relishes. Green tomatoes, peppers, cauliflower, and onions are common ingredients of mixed pickles, chowchow, etc.

Sauerkraut is an extremely important pickled product. Advertising of the healthfulness of its juice has increased the demand for "kraut." The olive is the most important fruit used in pickled form. Various fruits, but more particularly peaches, figs, and pears, are used in important quantities for sweet spiced pickles.

Cucumber Pickles

The processes of preparing and pickling cucumbers apply in some respects to other vegetables. Cucumbers for pickling should be slightly under ripe rather than fully mature. Cucumbers must be picked frequently as the vines blossom and bear over a period of several weeks.

The cucumbers deteriorate rapidly after picking and should be delivered to the factory or salting station as promptly as possible. Sorting to remove "nubbin" and "crook" cucumbers and to grade the cucumbers roughly for size is frequently done before salting. The larger cucumbers sorted out at this time may be barreled for preparation of dill pickles. Size grading can be dòne by diverging hardwood strips or pieces of pipe set in a sturdy frame.

Salting and Fermentation. Cucumbers must undergo a preliminary fermentation in brine before they are placed in vinegar. A number of different types of bacteria and yeasts are present during fermentation, but the predominant and desirable organisms are lactic acid bacteria.

Brine of 40° salometer to a depth of about 1 ft. is placed in the bottom of the vat to prevent bruising of the cucumbers as the vat is filled. After the vat is filled, brine is added to cover the cucumbers. Some packers add salt at this time, relying on osmosis to form brine, but filling the vat with 40° salometer brine is considered to give better results. Brine of any desired concentration can be made by diluting a saturated brine made by the Lixate method. A circular wooden head is then placed in the vat over the cucumbers and is held in place by 4- by 4-in. cross pieces secured at the ends by iron clamps. The cover is necessary to keep the cucumbers submerged in the brine, particularly during fermentation.

The concentration of the brine placed on the cucumbers is generally 40° salometer, i.e., approximately 10 per cent salt. Since cucumbers contain more than 90 per cent water, the brine rapidly decreases in concentration during fermentation and storage. A minimum concentration of 10 per cent salt (40° salometer) should be maintained during fermentation, to prevent the growth of putrefactive organisms. On the other hand, if the concentration greatly exceeds 10 per cent salt, the activity of the lactic acid organisms is greatly retarded. During the first week of fermentation it is customary to add salt to the vat each day, draw off the brine from the bottom of the vat and pump it over the top until the brine is of uniform concentration throughout the vat. Fabian (1944) recommends that a 40° salometer brine be applied initially and 9 lb. of salt added soon thereafter per 100 lb. of cucumbers. The brine is then increased 1,) salometer per week until 60° salometer is reached. In very hot weather the rate of increase per week may have to be 5 to 6° salometer. The brine may be pumped into a box filled with crushed rock salt and fitted with a screen bottom. The brine is circulated until the desired salometerdegree is attained.

Le Fevre also Fabian, found that the addition of about 1 per cent of sugar (preferably dextrose) to the brine greatly improves the character of the fermentation because some cucumbers are deficient in sugar and are liable to develop undesirable types of bacteria unless a small amount of sugar is added to the brine. It is best added after the fermentation is well under way.

The fermentation and curing process normally requires from 4 to 6 weeks, during which period the brine is maintained at about 10 per cent salt (40° salometer) by adding salt to the top of the tank and occasionally pumping over in order to make the concentration uniform. Too frequent pumping over is considered undesirable, however, as it may stimulate the growth of aerobic spoilage organisms. When fermentation is complete, the salt concentration of the brine is gradually increased and maintained at about 15 per cent salt (60° salometer).

During fermentation and curing, the cucumbers change in color from a bright green to an olive-green or yellowish-green color, and the flesh of a completely cured cucumber becomes translucent and no longer chalky white and opaque.

The cucumbers carry considerable numbers of yeasts, bacteria, and molds. Consequently, the brine is well inoculated naturally. During the early stages of fermentation, gas-producing organisms predominate. Subsequently gas formation nearly ceases, but lactic fermentation continues. .

Normally the brine attains about 0.6 to 0.8 per cent total acidity expressed as lactic, although considerably higher acidity is sometimes reached. Some alcohol is formed by yeasts. Bacteria form small amounts of acetic and propionic acids.

The lactic acid bacteria occurring in pickle brines, when observed microscopically, are usually found to be long rods, frequently occurring in pairs. They resemble the so-called Tourne bacteria that cause spoiling of cider-vinegar stock and wine. The lactic bacteria are facultative anaerobes, growing better in the absence of air. They are more resistant to salt than are most spoilage bacteria; hence if the brine concentration is maintained at 40° salometer during fermentation, there is little danger of spoilage from these latter organisms.

In filling the vat the cucumbers and salt should be weighed and the water metered in order that the resulting brine may be of the desired concentration: One pound of salt per gallon makes a brine of approximately 41° salometer. At one time dry salting was generally used. Dry salt was added and water was extracted by osmosis from the cucumbers to form a brine. The objection to this method is that it may cause collapse of the cucumbers to such an extent that they do not regain their original size and form during processing after removal from the brine.

During fermentation the cucumbers become very permeable and consequently absorb salt rapidly to come to equilibrium with the surrounding brine.

Other Vinegar Pickles and Relishes

Many kinds of fruit and vegetable pickles are produced commercially.

Onions. Small onions are first trimmed and peeled. They are generally stored in several changes of water for 3 or 4 days and are then, in some pickling plants, placed in brine strong enough to prevent fermentation, i.e., about 600 (15 per cent salt), and stored until they have become translucent or until used for pickles. The brine is strengthened by addition of salt as required. The salt is leached from the onions with warm water before they are placed in vinegar. Onions are also prepared for pickling by fermentation in brine of 10 per cent salt (40° salometer), as described for cucumbers earlier in this chapter.

Green Tomatoes and Mango Peppers. These are usually handled in the same manner as cucumbers, but in strong brine in order to minimize gaseous fermentation.

Cauliflower. Cauliflower in some factories is placed at once in a strong brine, 60° salometer, and fermentation is prevented by maintaining the brine at this concentration until the cauliflower is cured. Le Fevre, however, recommends that cauliflower be cured in a 10 per cent brine (40° salometer) and prepared for the vinegar in the same manner as cucumbers.

String Beans. These are usually cured in barrels after mixing with about 60 lb. of salt per 50-gal. barrel. The salt withdraws juice from the beans to give a strong brine. The beans may also be cured by fermentation in brine in the same manner as cucumbers.

Small Peppers. Small peppers for Tabasco sauce, etc., are fermented in wood in brine of about the same concentration as that used for cucumbers. They are also packed fresh in vinegar or they are barreled directly in strong vinegar containing some added salt, the purpose being to prevent .gaseous spoilage.

Processing and Addition of Vinegar. These various vegetables after curing in brine are prepared and stored in vinegar in much the same manner as described elsewhere for cucumber pickles.

Sweet Fruit Pickles. Peaches, pears, figs, watermelon rind, and grapes are often prepared as sweet pickles. The following method is satisfactory. The fruit is cooked in water or in dilute syrup until tender, then boiled a short time in a syrup of sugar 24 lb., water 2 gal, vinegar 1 gal, and 10 oz. each of whole cloves, stick cinnamon, and ginger; and allowed to stand overnight. The syrup is drawn off and concentrated to a boiling point of about.

Chapter 3

Fruits and Vegetable Processing

Fruit Nutrition Facts

Why Fruits

Fruits are nature's wonderful medicines packed with vitamins, minerals, anti-oxidants and many phyto-nutrients (Plant derived micronutrients). They are absolute feast to our sight, not just because of their color and flavor but help body keep fit and healthy.

1. Fruits are low in calories and fat and are a source of simple sugars, fiber, and vitamins, which are essential for optimizing our health.
2. Fruits provide plenty of soluble dietary fiber, which helps to ward of cholesterol and fats from the body and to get relief from constipation as well.
3. Fruits contain many anti-oxidants like poly-phenolic flavonoids, vitamin-C, anthocyanins. These compounds, firstly, help body protect from oxidant stress, diseases, and cancers, and secondly, help body develop capacity to fight against these ailments by boosting our immunity level. Many fruits, when compared to vegetables and cereals, have very high anti-oxidant values which is something measured by their "Oxygen Radical Absorbent Capacity" or ORAC.
4. Anthocyanins are flavonoid category of poly-phenolic compounds found in some "blue fruits" like blue-black grapes, mulberries, acai berry, chokeberries, blueberries, blackberries, and in many vegetables featuring blue or deep purple color. Eating fruits rich in blue pigments offers many health benefits. These compounds have potent anti-oxidant properties, remove free radicals from the body, and thus offer protection against

cancers, aging, infections etc. These pigments tend to concentrate just underneath the skin.

5. Fruit's health benefiting properties are because of their richness in vitamins, minerals, micro-nutrients, anti-oxidants which helps body prevent or at least prolong the natural changes of aging by protecting and rejuvenating cells, tissues and organs in the human body. The overall benefits are manifold. Fruit nutrition benefits are infinite. You are protecting yourself from minor problems like wrinkling of skin, hair fall, memory loss to major ailments like age related macular degeneration of retina in the eyes, Alzheimer's disease, colon cancers, weak bones (osteoporosis)... and the list of fruit nutrition benefits never ends

Fruit and Vegetable Processing

Fruits and vegetables are generally used fresh and a large proportion is processed also. It is very perishable in nature, so it is required to be consumed and/ or processed as early as possible. Fruits and vegetables are washed, blanched and peeled for canning, freezing and drying. Washing, blanching, and peeling fruits and vegetables for canning, freezing, or drying may be conveniently considered together, since in many instances all the operations are conducted simultaneously.

The fruits and vegetables are washed with clean water. Water is used for five purposes

- ➢ for the generation of the steam used in sterilizing,
- ➢ for lye peeling, washing, and other preliminary treatment of the raw materials,
- ➢ for the preparation of brine and syrup,
- ➢ for washing the floors, machinery, and cans, and
- ➢ for cooling the canned product.

Soaking—Fruits or vegetables may be washed with water in three different ways: soaking, washing by agitation, and sprays. Soaking is not in itself an effective means of removing dirt but is useful as a preliminary treatment to washing by sprays or by agitation. Hot water is a more effective soaking agent than cold water. It is essential that the water is abundant and changed frequently; otherwise the soaking vat may serve as a source of contamination rather than as a means of cleansing. Wash water is often continuously chlorinated.

Washing by Agitating in Water—If the fruits or vegetables are agitated in water, the efficiency of the soaking process is greatly enhanced. Compressed air is sometimes used to agitate water in tanks in which the fruit or vegetables are to be washed, as in one method of washing spinach, or it may also be agitated or circulated by means of a pump. In the washing of some vegetables and oranges a detergent is added to improve cleansing action.

Washing by Sprays—Washing of fruit and vegetables by means of sprays of water is by far the most satisfactory method. Products that are heavily contaminated with soil or other objectionable material should first be soaked thoroughly to loosen adhering soil before washing under sprays. The spray in which a small volume of water under heavy pressure is used is very much more effective than the one in which a large volume of water under low pressure is employed.

Scalding and Blanching

Most vegetables are heated in water or in live steam before canning, this treatment being known among canners as "blanching." It cleanses the product and decreases its volume so that a well-filled can is obtained; in some cases it removes disagreeable odors or flavors; and with some vegetables it removes slime-forming substances. It may or may not aid in retention of the green color of the vegetables, depending upon the vegetable, the temperature used in blanching, and the method of preservation used after blanching.

The severity of blanching should be regulated according to the maturity and tenderness of the vegetable by varying the length of the blanching period or the temperature of the water. Hard water that is high in calcium or magnesium salts will cause hardening and toughening of peas blanched in it, probably owing to reaction with pectic substances. In blanching green vegetables for freezing storage it is necessary to heat sufficiently to destroy the catalase and peroxidase enzyme in order that the frozen vegetables will not develop a hay-like odor and flavor in storage.

Corn, cream-style for canning, is not blanched on the cob but it is given a precook before canning; but that for freezing is blanched on the cob. Tomatoes are blanched in steam or boiling water for a short time to crack and loosen the skins. Sweet potatoes and beets are usually heated in steam or in steam under pressure in order to facilitate peeling.

Peeling Fruits and Vegetables for Canning

The quality of certain canned fruits and vegetables depends in large measure upon the care taken in peeling.

Hand Peeling – Initially peeling was done by hand. Presently the lye-peeling system is used. Four different methods of peeling are in use.

1. In one process the pimientos are passed through a revolving steel cylinder that is heated by a direct gas flame. The pimientos are roasted, and the peels may be easily removed from the roasted product by hand or heavy sprays.

2. The second process consists in passing the pimientos through a bath of cottonseed oil at about 400°F. This accomplishes the same purpose as the roasting process.
3. The third process of peeling pimientos is with a dilute boiling lye solution.
4. The fourth method is that of heating for a short time to a high temperature after treatment in strong lye solution generally used for beets, carrots, and apple.

Mechanical Peeling

Apples are peeled in machines that remove the peeling, core the fruit and cut it in circular slices. Root vegetables such as carrots, turnips, parsnips, etc., can be peeled in a mechanical peeler consisting of an upright cylinder provided in the bottom with a rapidly revolving disk, which in addition to its rotary motion undergoes an undulatory movement.

Lye Peeling

Lye peeling was probably first used commercially in the production of hominy, when the corn was peeled with lye consisting of the leachings from wood ashes. Usually a small amount of sodium hydroxide remains in the finished product. Lye peeling is used on peaches, sweet potatoes, apricots, and carrots.

Heat Peeling

Placing the halved or whole fruit on trays in a steam box can peel fruits for 2 or 3 min. or by immersing the halved fruit in boiling water for a short time. This treatment loosens the skins so that they may be easily slipped from the fruit with the hands. Tomatoes are blanched in steam or in boiling water, and then immersed or sprayed with cold water to cool the fruit and to loosen the skins. After blanching and cooling, the tomatoes are easily peeled by hand. Boiling water is probably more desirable than steam as a blanching agent for tomatoes for the reason that it heats them uniformly and cleanses them in addition to loosening the skins. The time for immersion in boiling water is approximately 30 to 60 sec.

Sodium hydroxide is the most common lye used in the peeling of fruits. A mixture of sodium carbonate and sodium hydroxide may also be used, although the action of the carbonate is much less vigorous than the action of the hydroxide. However, the presence of the carbonate makes it much less difficult to wash the lye from the fruit. The sodium carbonate-sodium hydroxide mixture is sold as "canners' alkali."

There are advantages of lye peeling in comparison to the hand peeling

- reduced cost of peeling,

- more rapid handling of the fruit, and
- less loss of fruit by peeling.

The length of immersion varies from about 2 to about 5 min. The time of immersion or the concentration of the lye solution can be varied to suit the condition of the fruit. Generally the consumption of lye is 5 to 8 lb. per ton of fresh fruit.

Grading Fruits and Vegetables for Canning and Freezing

One of the most important factors in determining the quality of processed fruit and vegetable products is careful grading. Various attributes of the raw materials and of the processed products are used as criteria in grading. Size, color, maturity, freedom from blemishes or other defects, symmetry, texture, freedom from damage by mold or insects, flavor, and odor are examples of properties that are taken into account in grading or judging the raw or the finished product.

Size Grading of Fruits

Cherries, plums, grapes, berries, and olives are graded for size whole, while peaches and apricots are graded after cutting in half or after halving and peeling. Pears are graded for size mechanically before peeling or halving.

Syrups and Brines Used in Canning

In canning, syrups are added to fruits and brines are added to vegetables to improve the flavor, fill the spaces between the pieces of canned product, and aid in the transfer of heat during processing.

Sugars

The principal sugar used in canning is sucrose, i.e., cane or beet sugar. Dextrose (corn sugar) is also important in fruit canning. Cane and beet sugars are identical in chemical composition and are used interchangeably in commercial canning. Unrefined sugars may occasionally contain sulfur dioxide, which may form hydrogen sulfide in the cans and a black deposit of metallic sulfide. It also hastens the corrosion of cans.

Invert Sugar

The principal sugar in all these commercial sugars is sucrose, $C_{12}H_{22}O_{11}$, which on hydrolysis yields equal amounts of glucose and fructose. The hydrolyzed product is known as "invertsugar." It is considerably sweeter than sucrose. For some purposes sucrose sugar is dissolved in water, inverted with citric acid or with invertase, and concentrated to heavy syrup, used extensively in candy making. Corn sugar, or glucose, in the form of a heavy syrup containing dextrins, is sometimes used in the manufacture of low-priced jams and Jellies. It is very much less sweet than cane sugar, is produced from cornstarch by hydrolysis under pressure with dilute hydrochloric or sulphuric acid, and possesses the chemical formula $C_6H_{12}O_6$.

Processing

"Processing," i.e.,-heating or sterilization of canned foods, is the most important operation in canning. The canned product often is not sterile in a bacteriological sense because of the presence of heat-resistant aerobes or thermophiles that do not develop in the canned product because of unfavorable conditions. Its principal purposes are (1) to render the product stable against spoilage by microorganisms and (2) to improve the texture, flavor, and appearance by cooking.

Principles of Processing

Time and Temperature - In the processing of canned foods both time and temperature are important, since as the temperature increases, the sterilizing effect is rapidly increased thus sterilization at 250°F is approximately 100 times as rapid as at 212°F.

Effect of Condition of Raw Material on Sterilization-Overripe fruit, because of the fact that it softens and forms a compact mass in the can and retards heat penetration, is more difficult to sterilize than fruit of firmer texture, which retains its normal shape and texture. Also, its low acidity makes sterilization more difficult.

A fruit is the edible, and more or less juicy, product of a tree or plant and consist of the matured ovary including its seeds and adjacent parts. Usually fruits are sweet, with a wide range of flavours, colours and textures.

Composition

Fruits are very poor source of protein and fat. Avocado is the exception containing 28% fat. Fruits contain high amount of moisture hence they are highly perishable. They are also good source of fibre. Fruits are not very good sources of calories. Fruits like bananas give fairly good amount of calories. Ripe fruit contains a higher percentage of sugar than unripe fruit does and the sugar is in the form of sucrose, fructose and glucose. Generally fruits are poor source of iron, Mangoes are excellent source of carotenes. Oranges are fairly good source of beta carotene. Guavas are the best source of vitamin C. Citrus fruits are good source of vitamin C. Cashew fruits are inexpensive and rich in vitamin C, although there is variation of vitamin content from fruit to fruit, most fruits in the raw state contain some ascorbic acid. If fruits are bruised, peeled, cooked or exposed to air, large amounts of the vitamin may be oxidized. Apples are not only expensive; they contribute little to the nutritive value. They give fibre to the diet.

Value Added products from Fruits

Juice, RTS, nectar, squash, cordial, jam, preserve, toffee, amchur, pickle, chutney, canned product, fruit powder, concentrate, jelly, cheese, toffee, vinegar, syrup, candy, wine, dried product, marmalade, cider, pickle etc are the major ralue added products of fruits. The major division is as follows :

- Fruit Beverage
- Jam, Jelly, Marmalade
- Candy
- Preserve
- Dehydration Of Fruits
- Pickle Production

Fruit Beverages

Fruit beverages are easily digestible, highly refreshing, thirst quenching, appetizing and nutritionally far superior to many synthetic and aerated drinks. They can be classified into unfermented and fermented beverages.

Unfermented Beverages

Fruit juices which do not undergo alcoholic fermentation are termed as unfermented beverages. They include natural and sweetened juices, RTS, nectar, cordial, squash, crush, syrup, fruit juice concentrate and fruit juice powder. Barley waters and carbonated beverages are also included in this group.

A. Squash

This is a type of fruit beverage containing at least 25 per cent fruit juice or pulp and 40 to 50 per cent total soluble solids, commercially. It also contains about 1.0 per cent acid and 350 ppm sulphur dioxide or 600 ppm sodium benzoate. It is diluted before serving.

Mango, orange and pineapple are used for making squash commercially. It can also be prepared from lemon, bael, papaya, etc., using potassium metabisulphite (KMS) as preservative or from Jamun, passion-fruit, peach, plum, raspberry, strawberry, grapefruit etc., with sodium benzoate as preservative.

Flowchart for Processing of Squash

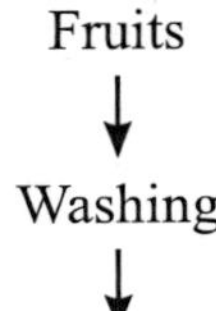

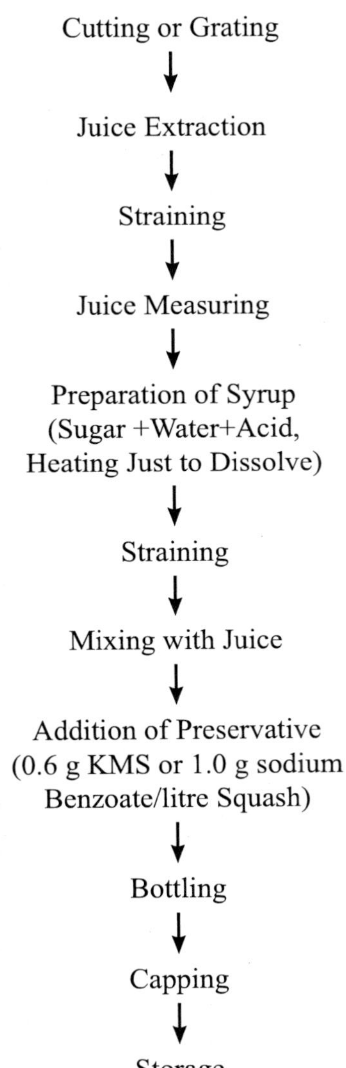

B. Ready-to-Serve (RTS)

This is a type of fruit beverage which contains at least 10 per cent fruit juice and 10 per cent total soluble solids besides about 0.3 per cent acid. It is not diluted before serving, hence it is known as ready-to-serve (RTS).

Flow-Sheet for Processing of RTS Beverages

Fruits
(Pulp/Juice)
↓
Mixing with Strained Syrup Solution (Sugar + Water + Acid, Heated Just to Dissolve) According to Recipe
↓
Homogenization
↓
Bottling
↓
Crown Corking
↓
Pasteurization
(at about 90ºC) for 25 Minutes
↓
Cooling
↓
Storage

C. Cordial

It is a sparkling, clear, sweetened fruit juice from which pulp and other insoluble substances have been completely removed. It contains at least 25 per cent juice and 30 per cent TSS. It also contains about 1.5 per cent acid and 350 ppm of sulphur dioxide. This is very suitable for blending with wines. Lime and lemon are suitable for making cordial.

Flow Chart for Processing of Cordial

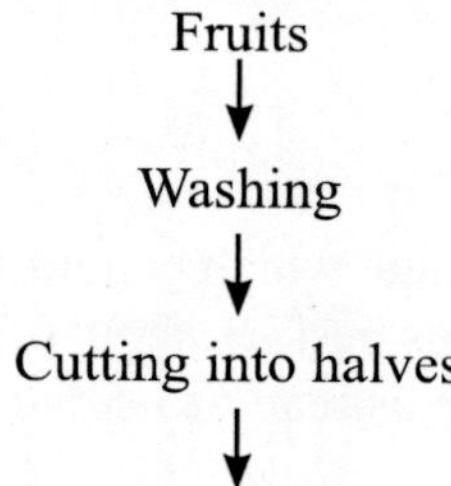

Fruits
↓
Washing
↓
Cutting into halves
↓

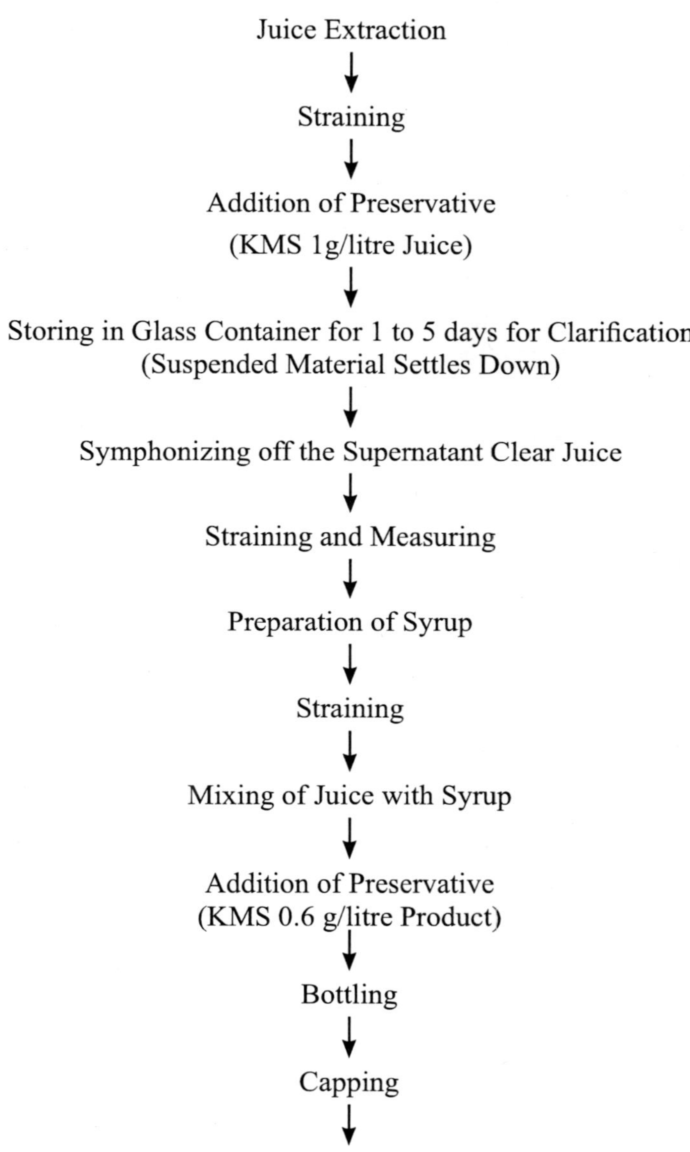

D. Nectar

This type of fruit beverage contains at least 20 per cent fruit juice/pulp and 15 per cent total soluble solids and also about 0.3 per cent acid. It is not diluted before serving.

S.N.	*Fuit*	*Juice/Pulp (%)*	*Quantity of Water Required (litre)*
1.	Mango	20	Quantity of Finished Product (litre) - Quantity of (Juice (litre) + Sugar (kg) + Acid (kg) Used
2.	Papaya	20	
3.	Guava	20	
4.	Bael	20	
5.	Jamun	20	
6.	Aonla (Blend)	Aonla Pulp 20 Lime Juice2 Ginger Juice 1	

For preparing the above beverages, the total soluble solids and total acid present in the pulp/juice are first determined and then the requisite amounts of sugar and citric acid dissolved in water are added for adjustment of TSS and acidity.

Fermented Beverages—Fruit juices which have undergone alcoholic fermentation by yeasts include wine, champagne, port, sherry, tokay, muscat, perry, orange wine, berry wine, nira and cider.

Flow Chart for Processing of Wine

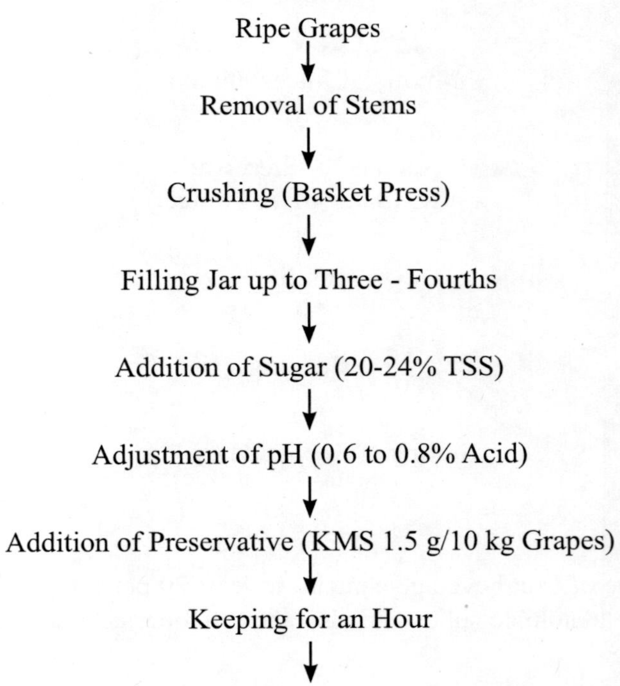

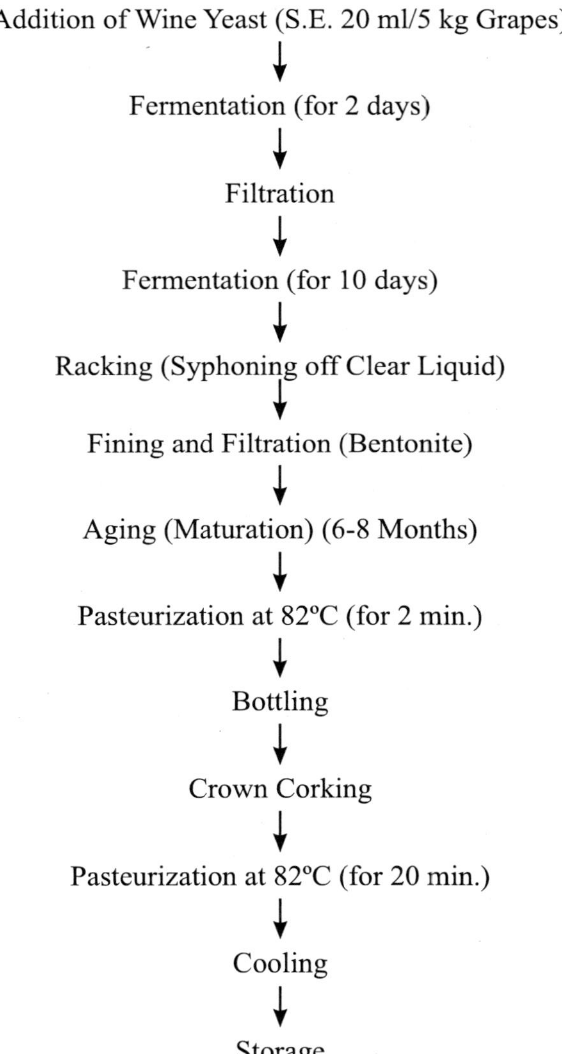

II. Jam, Jelly, Marmalade

A. Jam

Jam is a product made by boiling fruit pulp with sufficient amount of sugar to a reasonably thick consistency, firm enough to hold the fruit tissues in position. Apple, Pear, Sapota (chiku), peach, papaya, karonda, carrot, plum, straw-berry, raspberry, mango, tomato, grapes and muskmelon are used for preparation of jams. It can be prepared from one kind of fruit or from two or more kinds.

Flow Chart for Processing of Jam

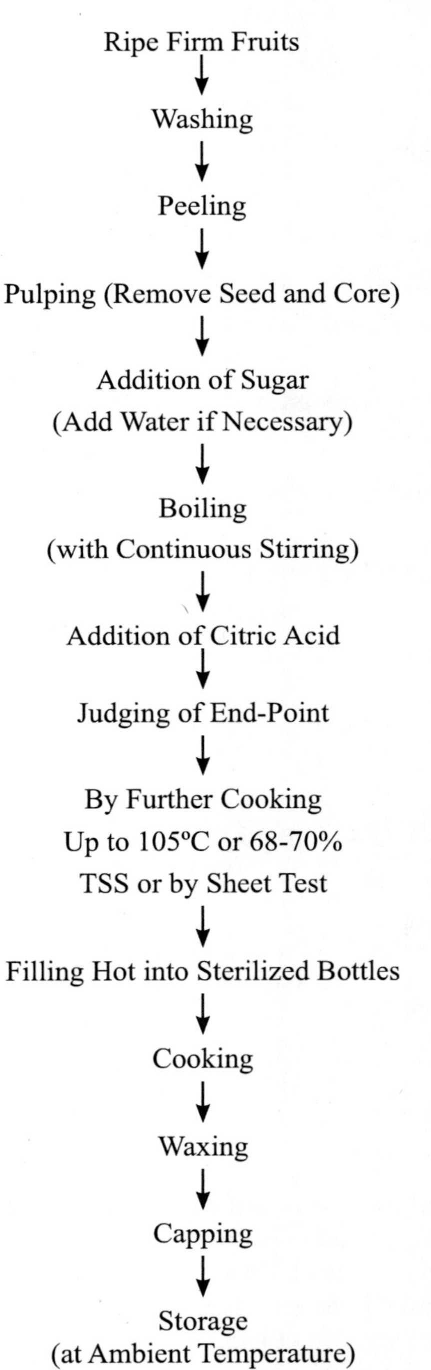

B. Jelly

A jelly is a semi-solid product prepared by boiling a clear, strained solution of pectin-containing fruit extract, free from pulp, after the addition of sugar and acid. A perfect jelly should be transparent, well-set, but not too stiff, and should have the original flavour of the fruit. It should be of attractive colour and keep its shape when removed from the mould. It should be firm enough to retain a sharp edge but tender enough to quiver when pressed.

Guava, sour apple, plum, karonda, wood apple, loquat, papaya and goose-berry are generally used for preparation of jelly. Apricot, pineapple, strawberry, raspberry, etc., can be used but only after addition of pectin powder, because these fruits have low pectin content.

Flow Chart for Processing of Jelly

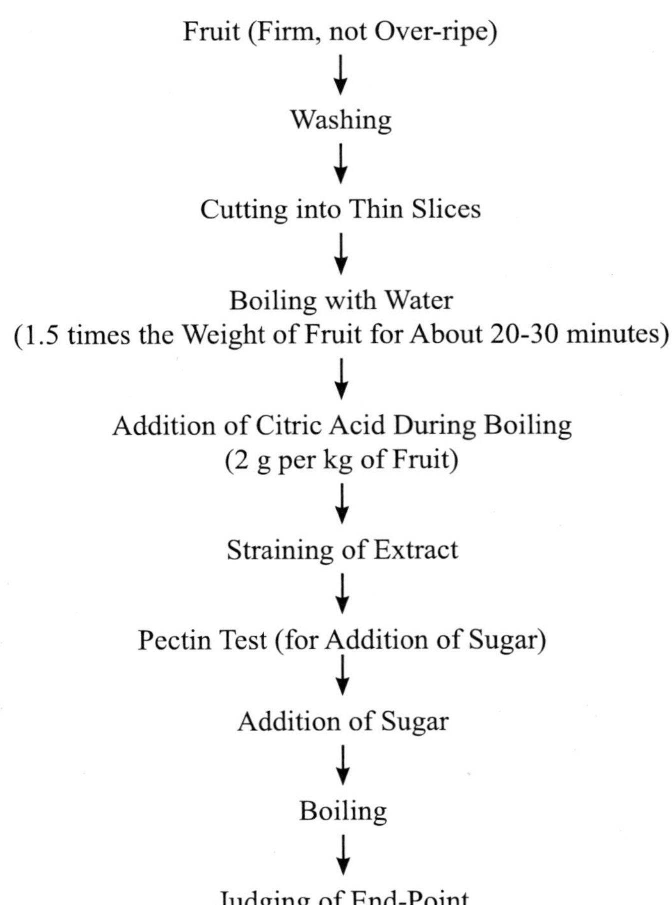

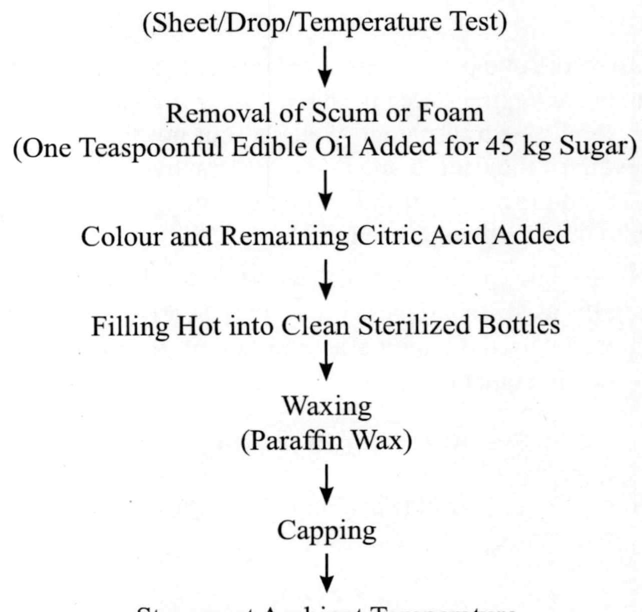

C. Marmalade

This is a fruit jelly in which slices of the fruit or its peel are suspended. The term is generally used for products made from citrus fruits like oranges and lemons in which shredded peel is used as the suspended material. Citrus marmalades are classified into (*i*) jelly marmalade, and (*ii*) jam marmalade.

1. ***Jelly Marmalade :*** The following combinations give good quality of jelly marmalade:-

(*i*) Sweet orange (Malta) and khatta or sour orange (Citrus aurantium) in the ratio of 2:1 by weight. Shreds of Malta orange peel are used.

(*ii*) Mandarin orange and khatta in the ratio of 2:1 by weight. Shreds of Malta orange peel are used.

(*iii*) Sweet orange (Malta) and galgal (**Citrus limonia**) in the ratio of 2:1 by weight. Shreds of Malta orange peel are used.

2. ***Jam Marmalade*** **:** The method of preparation is practically the same as that for jelly marmalade. In this case the pectin extract of fruit is not clarified and the whole pulp is used. Sugar is added according to the weight of fruit, generally in the proportion of 1:1. The pulp-sugar mixture is cooked till the TSS content reaches 65 per cent.

Flow Chart for Processing of Marmalade

Fruit
(Firm, not Over-ripe)

↓

Washing

↓

Cutting into Thin Slices

↓

Boiling with Water
(1.5 times the Weight of Fruit
for About 20-30 minutes)

↓

Addition of Citric Acid During Boiling
(2 g per kg of Fruit)

↓

Straining of Extract

↓

Pectin Test
(For Addition of Sugar)

↓

Addition of Sugar

↓

Boiling

↓

Judging of End-Point
(Sheet/Drop/Temperature Test)

↓

Removal of Scum or Foam
(One Teaspoonful Edible Oil
Added for 45 kg Sugar)

↓

Colour and Remaining Citric Acid Added

↓

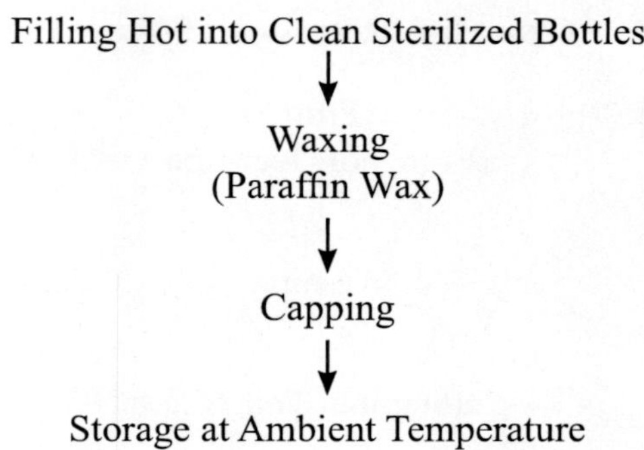

III. Candy

A whole fruit/vegetable or its pieces impregnated with cane sugar or glucose syrup, and subsequently drained free of syrup and dried, is known as candied fruit/vegetable. The most suitable fruits for candying are aonla, karonda, pineapple, cherry, papaya, apple, peach, and peels of orange, lemon, grapefruit and citron, ginger, etc.

The process for making candied fruit is practically similar to that for preserves. The only difference is that the fruit is impregnated with syrup having a higher percentage of sugar or glucose. A certain amount (25-30 per cent) of invert sugar or glucose, viz., confectioners glucose (corn syrup, crystal syrup or commercial glucose), dextrose or invert sugar is substituted for cane sugar. The total sugar content of the impregnated fruit is kept at about 75 per cent to prevent fermentation. The syrup left over from the candying process can be used for candying another batch of the same kind of fruit after suitable dilution for sweetening chutneys, sauces and pickles and in vinegar making.

Glazed Candy

Covering of candied fruits/vegetables with a thin transparent coating of sugar, which imparts them a glossy appearance, is known as glazing.

Cane sugar and water (2:1 by weight) are boiled in a steam pan at 113-114°C and the scum is removed as it comes up. Thereafter the syrup is cooled to 93°C and rubbed with a wooden ladle on the side of the pan when granulated sugar is obtained. Dried candied fruits are passed through this granulated portion of the sugar solution, one by one, by means of a fork, and then placed on trays in a warm dry room. They may also be dried in a drier at 49°C for 2-3 hours. When they become crisp, they are packed in airtight containers for storage.

Crystallized Candy

Candied fruits/vegetables when covered or coated with crystals of sugar, either by rolling in finely powdered sugar or by allowing sugar crystals to deposit on them from a dense syrup are called crystallized fruits. The candied fruits are placed on a wire mesh tray which is placed in a deep vessel. Cooled syrup (70 per cent total soluble solids) is gently poured over the fruit so as to cover it entirely. The whole mass is left undisturbed for 12 to 18 hours during which a thin coating of crystallized sugar is formed. The tray is then taken out carefully from the vessel and the surplus syrup drained off. The fruits are then placed in a single layer on wire mesh trays and dried at room temperature or at about 49°C in driers.

IV. Preserve

A mature fruit/vegetable or its pieces impregnated with heavy sugar syrup till it becomes tender and transparent is known as a preserve. Aonla, bael, apple, pear, mango, cherry, karonda, strawberry, pineapple, papaya, etc., can be used for making preserves.

Intermediate-moisture foods or semi moist foods, in one form or another, have been important items of diet for a very long time. Generally, they contain moderate levels of moisture, of the order of 20-50% by weight, which is less than is normally present in natural fruits and vegetables, but more than is left in conventionally dehydrated products. In addition, intermediate-moisture foods contain sufficient dissolved solutes to decrease water activity below that required to support microbial growth. As a consequence, intermediate-moisture foods do not require refrigeration to prevent microbial deterioration. There are various kinds of intermediate-moisture foods : natural products such as honey; manufactured confectionery product high in sugar, jellies, jams, and bakery items such as fruit cakes; and partially dried products including figs, dates, etc. In all of these products, preservation is partially from high osmotic pressure associated with the high concentration of solutes; in some, additional preservative effect is contributed by salt, acid and other specific solutes.

V. Dehydration of Fruits

Dehydration means the process of removal of moisture by the application of artificial heat under controlled conditions of temperature, humidity and air flow. In this process a single layer of fruits, whole or cut into pieces or slices are spread on trays which are placed inside the dehydrator. The initial temperature of the dehydrator is usually 43°C which is gradually increased to 66-71°C for fruits.

Basic Types of Drying Process

➢ Sun drying and solar drying

- Atmospheric drying including batch (kiln, tower and cabinet driers) and continuous (tunnel, belt, belt-trough, fluidized bed, puff, foam-mat, spray, drum and microwave);
- Sub-atmospheric dehydration (vacuum shelf/belt and freeze driers).
- Sun and solar drying of fruits and vegetables is a cheap method of preservation because it uses the natural resource / source of heat; sunlight. This method can be used on a commercial scale as well as the village level provided that the climate is hot, relatively dry and free of rainfall during and immediately after the normal harvesting period.
- Large scale driers are more promising than small scale ones. However, small scale driers should not be neglected.
- The drier should be designed to maximize the utilization factor of the capital investment, i.e. multi-products (fruits, vegetables and other raw material) and multi-use (e.g. Drying and heating water for domestic use).
- In general, an auxiliary heat source should be provided to assure reliability, to handle peak loads and also to provide continuous drying during periods of no sunshine.

Shade Drying

Shade drying is carried out for products which can loose their colour and / or turn brown if put in direct sunlight. Therefore, shade drying is carried out under a roof which has open sides.

Osmotic Dehydration

In osmotic dehydration the prepared fresh material is soaked in a heavy (thick liquid sugar solution) and / or a strong salt solution and then the material is sun or solar dried.

Common Driers Used for Drying/Dehydration

a. ***Air Convection Driers***

1. Kiln drier
2. Cabinet, tray and pan driers
3. Tunnel and continuous belt driers
4. Belt trough drier
5. Air lift drier
6. Fluidized bed drier
7. Spray driers

b. ***Drum or Roller Driers***

c. ***Vacuum Driers***

1. Vacuum shelf driers
2. Continuous vacuum belt drier
3. Freeze-drying

Flow Sheet for Drying/Dehydration of Fruits

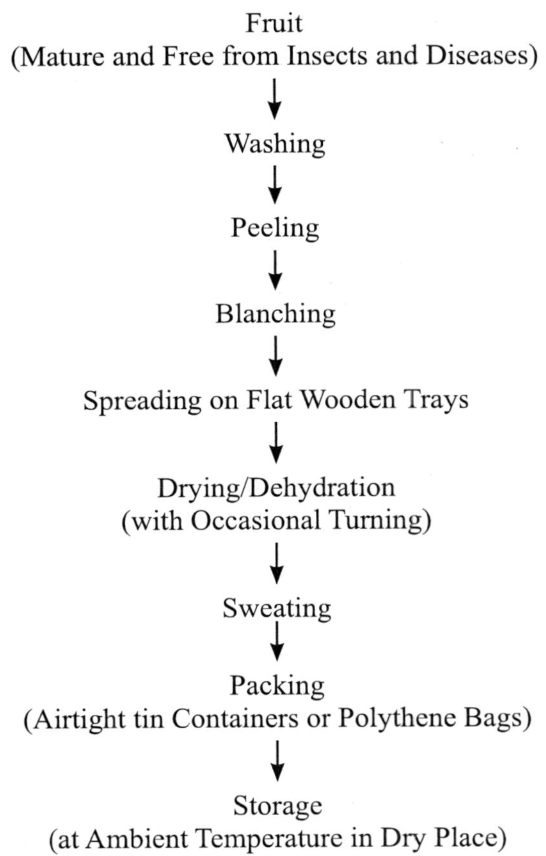

VI. Pickle Production

The preservation of food in common salt or in vinegar is known as pickling. It is one of the most ancient methods of preserving fruits and vegetables. Pickles are good appetizers and add to the palatability of a meal. They stimulate the flow of gastric juice and thus help in digestion. At present, pickles are prepared with salt, vinegar, oil or with a mixture of salt, oil, spices and vinegar.

Chapter 4

Fruits and Vegetables in India

India is the second largest producer of Fruits after China, with a production of 44.04 million tonnes of fruits from an area of 3.72 million hectares. A large variety of fruits are grown in India, of which mango, banana, citrus, guava, grape, pineapple and apple are the major ones. Apart from these, fruits like papaya, sapota, annona, phalsa, jackfruit, ber, pomegranate in tropical and sub-tropical group and peach, pear, almond, walnut, apricot and strawberry in the temperate group are also grown in a sizeable area. Although fruit is grown throughout the country, the major fruit growing states are Maharashtra, Tamil Nadu, Karnataka, Andhra Pradesh, Bihar, Uttar Pradesh and Gujarat.

Fruits

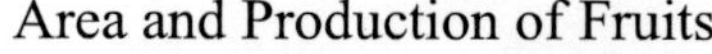

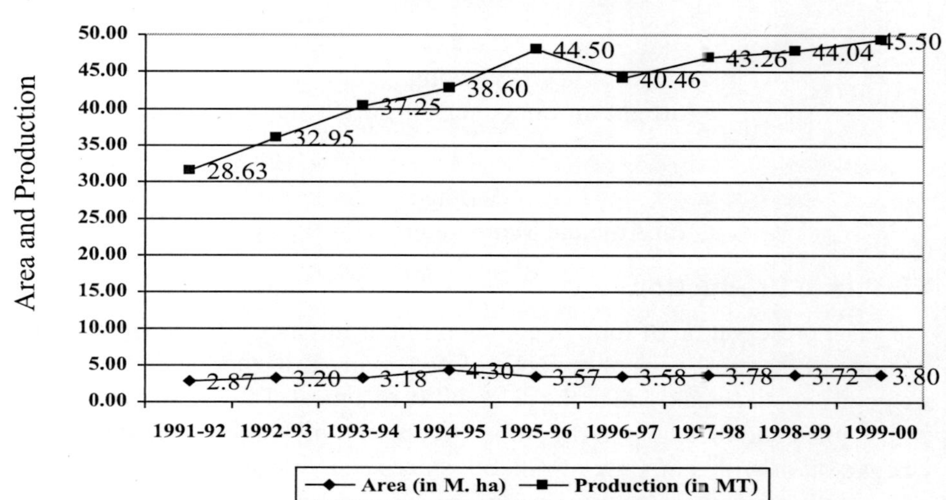

Percent Share of Fruits Production (1999-2000)

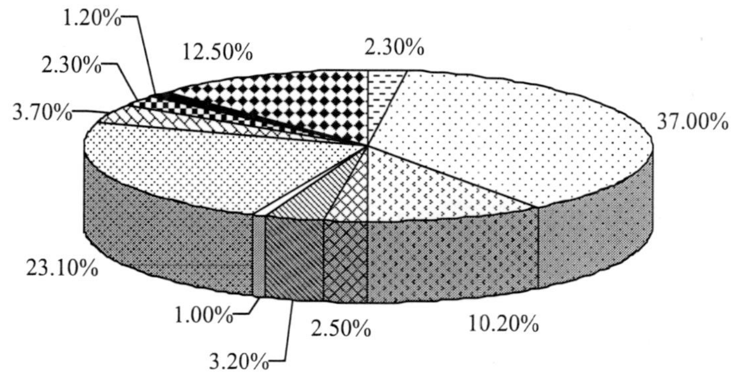

Area and Production of Fruits in India During 1999-2000

Crop	*Area (000 ha.)*	*Production (000 MT)*
Apple	238.3	1047.4
Banana	490.7	16813.5
Citrus	526.9	4650.6
Grapes	44.3	1137.8
Guava	150.9	1710.5
Litchi	56.4	433.2
Mango	1486.9	10503.5
Papaya	60.5	1666.2
Pineapple	75.5	1025.4
Sapota	64.4	800.3
Others	601.2	5707.6
Total	3796.8	45496.0

Indian Horticulture Production at a Glance (1991-92 and 2001-02 to 2010-11)

Year	Fruits		Vegetables		Flowers		Nuts	
	A	P	A	P	A	P (Loose)	A	P
1991-92	2874	28632	5593	58532	NA	NA	NA	NA
2001-02	4010	43001	6156	88622	106	535	117	114
2002-03	3788	45203	6092	84815	70	735	117	114
2003-04	4661	45942	6082	88334	101	580	106	121
2004-05	5049	50867	6744	101246	118	659	106	121
2005-06	5324	55356	7213	111399	129	654	130	149
2006-07	5554	59563	7581	114993	144	880	132	150
2007-08	5857	65587	7848	128449	166	868	132	177
2008-09	6101	68466	7981	129077	167	987	136	173
2009-10	6329	71516	7985	133738	183	1021	142	193
2010-11	6383	74878	8495	146554	191	1031	including in fruit	

Year	Aroma. & Medi. Crop		Plantation Crops		Spices		Mush-room	Honey	Grand Total	
	A	P	A	P	A	P	P	P	A	P
1991-92	NA	NA	2298	7498	2005	1900	NA	NA	12770	96562
2001-02	NA	NA	2984	9697	3220	3765	40	10	16592	145785
2002-03	NA	NA	2984	9697	3220	3765	40	10	16270	144380
2003-04	NA	NA	3102	13161	5155	5113	40	10	19208	153302
2004-05	131	159	3147	9835	3150	4001	40	10	18445	166939
2005-06	262	202	3283	11263	2366	3705	35	52	18707	182816
2006-07	324	178	3207	12007	2448	3953	37	51	19389	191813
2007-08	397	396	3190	11300	2617	4357	37	65	20207	211235
2008-09	430	430	3217	11336	2629	4145	37	65	20662	214716
2009-10	509	573	3265	11928	2464	4016	41	65	20876	223089
2010-11	510	605	3306	12007	2940	5350	41	65	21825	240531

Area and Production Estimates for Horticulture Crops
Area in 000' HA, Production in 000' MT and Productivity = MT/HA

	2008-09			2009-10			2010-11		
	Area	*Production*	*Pdy.*	*Area*	*Production*	*Pdy.*	*Area*	*Production*	*Pdy.*
Fruits									
Banana	709	26217	37.0	770	26470	34.4	830	29780	35.9
Mângo	2309	12750	5.5	2312	15027	6.5	2297	15188	6.6
Citrus	924	8623	9.3	987	9638	9.8	846	7464	8.8
Papaya	98	3629	37.1	96	3913	40.9	106	4196	39.6
Guava	204	2270	11.1	220	2572	11.7	205	2462	12.0
Apple	274	1985	7.2	283	1777	6.3	289	2891	10.0
Pineapple	84	1341	16.0	92	1387	15.1	89	1415	15.9
Sapota	156	1308	8.4	159	1347	8.5	160	1424	8.9
Grapes	80	1878	23.6	106	881	8.3	111	1235	11.1
Pomegranate	109	807	7.4	125	820	6.6	107	743	6.9
Litchi	72	423	5.9	74	483	6.5	78	497	6.4
Others	1083	7234	6.7	1105	7201	6.5	1265	7583	6.0
Fruits-Total	6101	68466	11.2	6329	7151	11.3	6383	74878	11.7
Vegetables									
Potato,	1828	34391	18.8	1835	36577	19.9	1863	42339	22.7
Tomato	599	11149	18.6	634	12433	19.6	865	16526	19.1
Onion	834	13565	16.3	756	12159	16.1	1604	15118	14.2
Brinjal	600	10378	17.3	590	10165	17.2	680	11896	17.5
Tapioca	280	9623	30.4	232	8060	34.8	221	8076	36.5
Cabbage	310	6870	22.1	331	7281	22.0	369	7949	21.5
Cauliflower	349	6532	18.7	338	6410	19.0	369	6745	18.3
Okra	432	4528	10.5	452	4803	10.6	498	5784	11.6
Peas	348	2916	8.4	365	3029	8.3	370	3517	9.5
Sweet Potato	124	1120	9.0	119	1095	9.2	113	1047	9.3
Others	2275	28006	12.3	2332	31724	13.6	2083	27557	13.2
Veg-Total	7981	129077	16.2	7985	133738	16.7	8495	146554	17.3
Aromatic	430	430	1.0	509	573	1.1	510	605	1.2
Almond/Walnut	136	173	1.3	142	193	1.4		(Including Fruits)	
Flowers loose	167	987		183	1021		191	1031	5.4
Flowers Cut*		47942			66671			69027	
Plantation Crops	3217	11336	3.5	3265	11928	3.7	3306	12007	3.6
Spices	2629	4145	1.6	2464	4016	1.6	2940	5350	1.8
Mushroom		37			41			41	
Honey		65			65			65	
Grand Total	**20662**	**214716**	**10.4**	**20876**	**223089**	**10.7**	**21825**	**240531**	**11.0**

NHB Data

Growth Trends

Crops	*08-09 Over 07-08*		*09-10 Over 08-09*		*10-11 Over 09-10*	
Horticulture	2.25	1.65	1.0	3.9	4.5	7.8
Fruit	4.16	4.39	3.7	4.5	0.8	4.7
Vegetable	1.69	0.49	0.1	3.6	6.4	9.6

*Lakh Numbers, not included in total and Area of Cut Flowers included in the area of loose flowers.

Totals may not match due to rounding off of figures

Crop-wise Area, Production and Productivity of Major Plantation Crops in India During 2008-09, 2009-10 and 2010-11

Area in 000 ha, Production in 000 MT and Productivity = MT/HA

	2008-09			**2009-10**			**2010-11**		
	Area	***Produc-tion***	***Pdy.***	***Area***	***Produc-tion***	***Pdy.***	***Area***	***Produc-tion***	***Pdy.***
Coconut	1903.2	10148.3	5.3	1895.2	10824.3	5.7	1895.9	10840.0	5.7
Cashewnut	893.0	695.0	0.8	923.0	613.0	0.7	953.2	674.6	0.7
Arecanut	387.1	481.3	1.2	400.1	478.0	1.2	400.1	478.0	1.2
Cocoa	34	11.8	0.3	46.3	12.9	0.3	56.5	14.4	0.3
Total	3217.3	11336.4	3.5	3264.6	11928.2	3.7	3305.7	12007.0	3.6

Crop-Wise Area, Production and Productivity of Major Spice Crops in India During 2008-09, 2009-10 AND 2010-11

Area in 000' HA, Production in 000' MT and Productivity = MT/HA

	2008-09			**2009-10**			**2010-11**		
	Area	***Produc-tion***	***Pdy.***	***Area***	***Produc-tion***	***Pdy.***	***Area***	***Produc-tion***	***Pdy.***
Chillies	779.05	1269.85	1.6	767.23	1202.94	1.6	792.1	1223.4	1.5
Garlic	166.21	831.10	5.0	164.86	833.97	5.1	200.6	1057.8	5.3
Turmeric	181.09	821.16	4.5	180.96	792.98	4.4	195.1	992.9	5.1
Ginger	108.64	380.10	3.5	107.54	385.33	3.6	149.1	702.0	4.7
Coriander	396.85	242.13	0.6	360.00	236.72	0.7	530.5	482.0	0.9

	2008-09			**2009-10**			**2010-11**		
	Area	***Production***	***Pdy.***	***Area***	***Production***	***Pdy.***	***Area***	***Production***	***Pdy.***
Tamarind	54.63	177.68	3.3	57.99	185.74	3.2	59.6	206.3	3.5
Cumin	42.938	172.47	0.4	377.01	156.33	0.4	507.8	314.2	0.6
Fenugreek	68.29	76.58	1.1	43.25	57.44	1.3	81.2	118.4	1.5
Fennel	47.16	64.29	1.4	50.67	56.55	1.1	61.8	105.4	1.7
Pepper	238.71	47.40	0.2	195.92	51.02	0.3	183.8	52.0	0.3
Cardamom	91.99	15.45	0.2	90.20	15.72	0.2	86.7	15.7	0.2
Other seed spices	17.71	10.71	0.6	19.36	13.04	0.7	0.0	0.0	0.0
Ajwan	19.59	16.41	0.8	17.06	10.27	0.6	25.8	22.2	0.9
Nutmeg	15.27	11.37	0.7	15.05	8.00	0.5	16.1	11.4	0.7
Tejpat/ Cinnamon	3.31	6.65	3.9	4.64	8.75	3.9	2.9	5.0	1.7
Clove	2.57	1.33	0.5	2.60	1.16	0.4	2.4	1.2	0.5
Others (2)	9.01	0.23	0.03	9.34	0.24	0.03	44.7	40.6	0.9
Total	2629.44	4144.91	1.6	2463.29	4015.91	1.6	2940.4	5350.5	1.8

NHB Data

Mango

Mango is the most important fruit covering about 35 per cent of area and accounting of 22 per cent total production of total fruits in the country, which is highest in the world with India's share of about 54%. India has the richest collection of mango cultivars. Major mango growing States are Uttar Pradesh, Bihar, Andhra Pradesh, Orissa, West Bengal, Maharashtra, Gujarat, Karnataka, Kerala and Tamil Nadu. The main varieties of mango grown in the country are Alphanso, Dashehari, Langra, Fajli, Chausa, Totapuri, and Neelum etc. Banana comes next in rank occupying about 13 per cent of the total area and accounting for about 34.2 per cent of the total production of fruits. India has first position in the world in banana production. While Tamil Nadu leads other States with a share of 19.00 per cent, Maharashtra has highest productivity of 58.60 metric tonnes against India's average of 32.50 metric tonnes per ha. The other major banana growing states are Karnataka, Gujarat, Andhra Pradesh and Assam The main varieties of banana are Dwarf Cavendish, Bhusaval Keli, Basrai, Poovan, Harichhal, Nendran, Safed velchi etc. Citrus fruits rank 3rd in area and production accounting for About 12 and 10.4 per cent of the total area and production respectively.

	Area (in 000' HA)				Production (in 000' MT)			
State/ UT's	**1996-97**	**1997-98**	**1998-99**	**1999-00**	**1996-97**	**1997-98**	**1998-99**	**1999-00**
Andhra Pradesh	406.7	414.5	378.6	449.2	5657.7	5899.1	4589.6	5175.4
Bihar	292.8	299.8	303.6	309.3	2752.2	3755.4	3797.2	3870.7
Gujrat	137.5	159.3	163	176.2	1820.0	2267.8	2293.5	2376.0
Karnataka	298.8	314.6	314.6	315.0	5133.6	5446.3	5446.3	5456.1
Kerala	195.8	195.8	233.1	187.8	1826.0	1826.0	1621.2	1184.5
Madhya Pradesh	59.5	62.4	63.1	67.4	1127.0	1184.0	1374.4	1536.1
Maharashtra	373.3	380.0	436.1	539.8	6333.1	6473.2	7521.7	8688.5
Orissa	206.6	227.4	249.4	204.9	1342.4	1511.8	1718.4	1202.9
Tamil Nadu	220.6	234.0	213.5	232.0	3862.7	3683.8	5447.6	5939.6
Uttar Pradesh (Hill)	185.5	186.7	187.9	172.5	510.2	515.3	520.4	475.5
Uttar Pradesh (Plain)	318.9	328.8	305.2	315.1	4045.1	4293.0	3097.8	3210.5
West Bengal	116.4	117.3	128.0	130.2	1035.1	1373.6	1536.0	1816.1
Others	767.0	781.3	750.7	697.4	5013.0	5033.9	5078.3	4564.1
Total	**3579.4**	**3701.9**	**3726.8**	**3796.8**	**40458.1**	**43263.2**	**44042.4**	**45496.0**

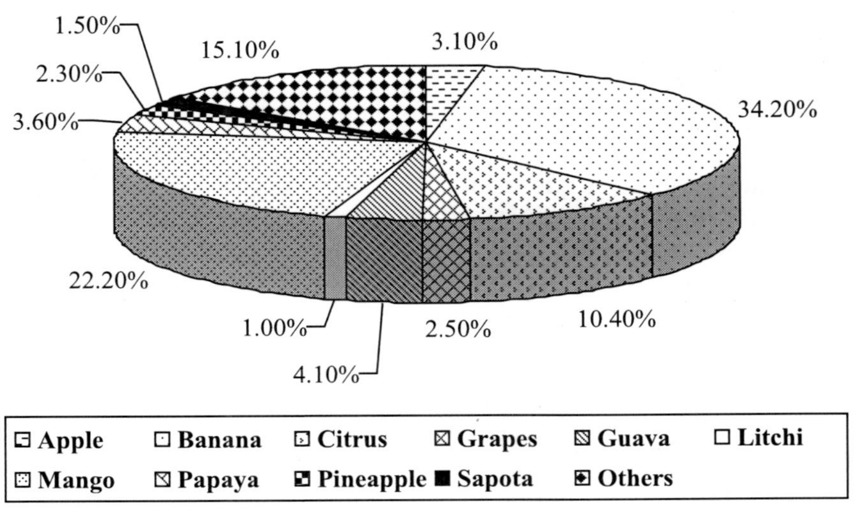
Production share of Major Fruits in India (1998-99)
1.50%
15.10%
3.10%
2.30%
34.20%
3.60%
22.20%
1.00%
2.50%
10.40%
4.10%
Apple
Banana
Citrus
Grapes
Guava
Litchi
Mango
Papaya
Pineapple
Sapota
Others

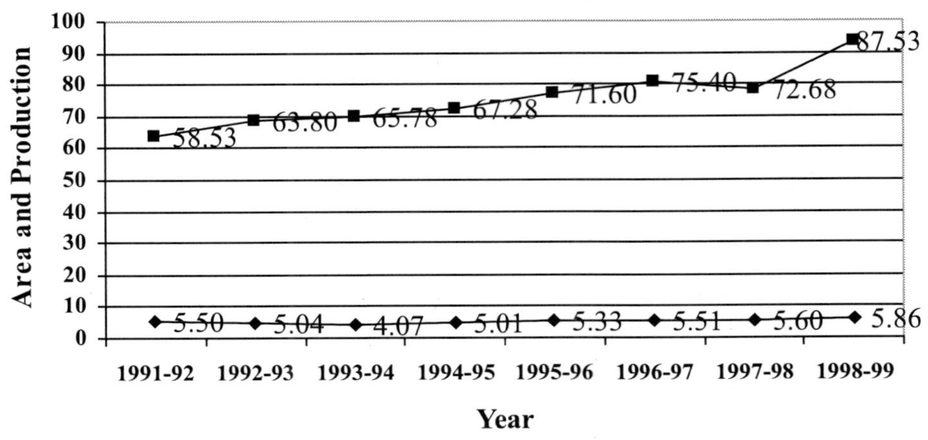
Production (In 000* MT)
Area (in 000 HA)
Area and Production of Vegetables
Area and Production
100
90
80
70
60
50
40
30
20
10
0
58.53
63.80
65.78
67.28
71.60
75.40
72.68
87.53
5.50
5.04
4.07
5.01
5.33
5.51
5.60
5.86
1991-92
1992-93
1993-94
1994-95
1995-96
1996-97
1997-98
1998-99
Year
Area (in M. ha)
Production (in MT)

Lime, lemons, sweet oranges and mandarin cover bulk of the area under these fruits and are grown mainly in Maharashtra, Andhra Pradesh, Karnataka, North-Eastern States, Punjab, Orissa and Madhya Pradesh. Guava is the fourth most widely grown fruit crop in India. The area under guava is about 0.15 Million ha producing 1.80 MT. The popular varieties of guava are Allahabad Safeda, Lucknoe-49, Nagpur Seedless, and Dharwar etc. Bihar is the leading state in guava production with 0.30 MT followed by Andhra Pradesh, and Uttar Pradesh. The other states where guava is grown widely are Gujarat, Karnataka, Punjab and Tamil Nadu. Grapes occupies fifth position amongst fruit crops with a production of 1.08 MT from an area of 0.04 Million ha. The major varieties of grapes grown in India are, Thomson Seedless, Sonaka, Anab-e-Shahi, Perlette, Banglore blue, Pusa seedless, Beauty seedless etc. Maharashtra occupies the first position with a production of 0.68 MT of grapes, followed by Karnataka. The other states growing grapes are Punjab, Andhra Pradesh and Tamil Nadu. The other major fruits grown in the country are Apple, Litchi, Papaya, Pineapple and Sapota.

In vegetables production, India is next only to China with an annual production of 87.53 million tonnes from 5.86 million hectares having a share of 14.4 per cent to the world production. Adoption of high yielding cultivars and FI hybrids and suitable production technologies has largely contributed for higher production and productivity. Per capital consumption has also increased from 95 gram to 175 gram per day. More than 40 kinds of vegetables belonging to different groups, namely cucurbits, cole crops, solanaceous, root and leafy vegetables, are grown in different agro-climatic situations of the country. Except a few, namely brinjal (egg plant), colocasia, cucumber, ridge gourd, sponge gourd, pointed gourd etc., most of the other vegetables have been introduced from abroad.

Potato is most widely grown vegetable crop in the country with a share of 25.7 per cent. The area under potato cultivation is 1.28 Million ha with total production of 22.49 MT. The main varieties of potato grown in the country are Kufri Chandramukhi, Kufri Jyoti, Kufri Badshah, Kufri Himalani, Kufri Sindhuri, Kufri Lalima etc. Uttar Pradesh is the leading potato growing state in the country with a production of 9.53 million tonnes followed by West Bengal and Bihar.

Tomato occupies second position amongst the vegetable crops in terms of production. The total production of tomato in the country in 1998-99 was 8.27 MT from an area of 0.46 M. ha. The main varieties of tomato grown in the country are Pusa Ruby, Pusa Early Dwarf, Arka Abha, Arka Alok, Pant Bahar, Pusa hybrid-1, Pusa hybrid-2, MTH-6, Arka Vardan etc. Andhra Pradesh is the largest grower of tomato with a production of 2.05 MT. The other main tomato growing states are Bihar, Karnataka, Maharashtra and Orissa.

Brinjal occupies the third position amongst vegetable crops. The production of brinjal in the year 1998-99 was 7.88 MT from an area 0.49 M.ha. The varieties

State-Wise Area and Production of vegetables

	Area (in 000' HA)				*Production (in 000' MT)*			
State / UT's	**1996-97**	**1997-98**	**1998-99**	**1999-00**	**1996-97**	**1997-98**	**1998-99**	**1999-00**
Andhra Pradesh	194.2	179.8	249.3	230.1	1895.0	2252.2	3541.2	2839.1
Assam	223.2	223.8	245.9	255.9	2074.1	2180.2	2834.8	3089.4
Bihar	602.4	603.6	616.6	626.0	8235.7	8266.2	9418.4	9548.8
Gujrat	171.5	153.0	189.9	201.0	2179.4	2176.9	3255.0	2647.0
Karnataka	312.1	294.8	309.7	361.6	4978.7	4944.9	4944.9	6796.9
Kerala	243.9	243.9	159.7	159.7	2790.0	2789.5	2857.2	2857.1
Madhya Pradesh	199.5	206.0	234.0	258.7	2889.5	2748.7	3276.2	3632.0
Maharashtra	352.3	276.0	341.2	385.3	4275.4	3317.2	4479.5	4828.6
Orissa	868.8	882.7	883.9	788.1	8746.0	9656.6	10087.1	9096.0
Tamil Nadu	164.4	177.0	206.7	209.1	3990.3	4085.4	5704.8	5660.3
Uttar Pradesh (Hill)	89.8	91.3	91.5	81.9	807.8	792.6	840.7	733.2
Uttar Pradesh (Plain)	647.8	638.2	640.7	688.9	12446.8	8623.4	12680.6	13842.4
West Bengal	831.7	1034.3	1100.0	1122.3	13670.8	15016.0	16367.4	17413.8
Others	613.8	602.7	596.9	624.3	6094.9	5833.3	7248.2	7846.1
Total	5515.4	5607.1	5866.0	5993.0	75074.4	72683.1	87536.0	90830.7

of brinjal popular in the country are Arka Navneet, Pusa Ankur, Hybrid-6, Pusa hybrid-5, ARBH-1, ABH-1, Pusa Purple Long, Pusa Purple Cluster, Ritu Raj etc. West Bengal is the largest producer of brinjal followed by Maharashtra and Bihar. The other main state growing brinjal are Karnataka, Maharashtra, Gujarat, Andhra Pradesh, Assam and Madhya Pradesh.

Cabbage is the fourth most widely grown vegetable crop of our country. India is the leading country producing Cabbage. The area under Cabbage cultivation is 0.23 M.ha producing 5.62 MT. The main varieties of cabbage are Pusa Drum Head, Golden Acre, Pride of India, Pusa Mukta, Pusa Synthetic etc. West Bengal produces 1.84 MT and is the largest grower of the cabbage. Orissa and Bihar occupies second and third position respectively. The other major growers of cabbage are Assam, Karnataka, Maharashtra and Gujarat. The other important vegetable crops grown in the country are onion, chillies, peas, beans, okra, cabbage, cauliflower, pumpkin, bottlegourd, cucumber, watermelon, palak, methi, carrot and radish.

SECTION TWO

FRUIT PROCESSING : PROCESS AND TECHNIQUES

Chapter 5

Aonla *(Emblica Officinalis Gaertn.)*

Aonla or Indian gooseberry *(Emblica Officinalis Gaertn.)* is known for its medicinal and therapeutic properties from the ancient time in India and considered as a wonder fruit for health conscious population. It has been grown and known in India for last more than 3500years. It belongs to family Euphorbiaceae. It is one of the important indigenous fruits of Indian subcontinent. In different parts of India, it is known by different vernacular names such as Amla or Aonla in Hindi, Dhatri, Dhatriphala, or Amalaki in Sanskrit, Amla or Amalaki in Bengali and Oriya, Nelli in Malayalam and Tamil Amlakamu in Telugu, Amolphal in Punjabi (Gurmukhi) and Aonla, Myrobalan or Indian gooseberry in English. Owing to its significant medicinal and nutritive value it is considered as Amrit Phal (life giving fruit).

Aonla is not only a source of nutrients and medicine, but cultivation of this crop is also highly remunerative for the farmers having marginal land. Traditionally, aonla has been a crop of forest or household, but during the last decade, there has been unprecedented expansion in the area under aonla cultivation across the country, utilizing the wasteland. This has resulted in efficient utilization of resources leading to better income to farmers, nutritional security coupled with enhanced employment and rehabilitation of wastelands.

Nutritional Value and Chemical Composition

Various parts of aonla tree are of great economic importance. The edible fruit tissue contains protein concentration 3-fold and ascorbic acid concentration 160-fold than those of apple. The fruit also contains considerably higher concentration of most minerals and amino acids than apple. Glutamic acid, proline, aspartic acid, alanine and lysine are 29.6, 14.6, 8.1, and 5.4 and 5.3%, respectively of the total amino acids.

The chemical composition of aonla fruit is presented in Table.

Table : Chemical Composition of Aonla Fruit

Constituent	*Amount*	*Constituent*	*Amount*
	(%)		*(mg/100g pulp)*
Moisture	77.1-82.20	Iron	1.20
Protein	0.50	Nicotinic acid	0.20
Fat	0.10	Vitamin C	200-1814
Minerals	0.5-0.70	Carotene	0.01
Fibre	1.9-34	Thiamine	0.03
Carbohydrate	14.10-21.89	Riboflavin	0.05
Niacin	0.18	Calcium	0.012-0.050
Tryptophan	3.00	Phosphorus	020-0.026
Methionine	2.00	Lysine	7.00

The pulpy portion of fruit is highly nutritive and is one of the richest sources of vitamin C. Aonla fruit is acidic, acrid, cooling, refrigerant, diuretic and laxative (Gopalan and Mohanram 1996). Normally one aonla fruit contains 20 times vitamin C in terms of antiscorbutic value than two oranges. The main constituents of aonla are tannins, polyphenolic compound, I, 3, 6- trigallcylglucose, terehebin, corilagin, phyllantine, beta siotosturol, linoleic acid, ellagic acid and lupeol. Aonla fruit contains a fair amount of polyphenols and tannins, which retard the oxidation of vitamin C. It is not only a rich but also a cheapest source of vitamin C and a fair source of carbohydrates, carotene, thiamine, riboflavin and minerals (iron, calcium and phosphorus). Aonla is a rare example of an edible material, which is rich in tannins as well as ascorbic acid (Kalra 1988).

Products

In general aonla fruits are utilized for the following three purposes (Singh et al. 1993). These products are: (*i*) food items (*ii*) cosmetic preparations and (*iii*) ayurvedic preparations. Here we'll discuss the food items prepared from aonla.

Food items

Processed products of high quality can only be made from good quality of raw materials. Amongst the aonla cultivars, NA-6 and Krishna (NA-5) have low fiber content and phenols with average composition of vitamin C and minerals. Hence, these have better potentiality for their use in the processing industry particularly for preserve, candy and shred making. Owing to better recovery and smaller fruit size, Kanchan (NA-4) and NA-7 have better demand for other preparations (Singh and Pathak 1987; Tondon et al. 2003). Steps involved in preparation of food items are enumerated below.

Pulp extraction technique

Extraction of pulp from aonla fruit is necessary for processing of various products. Fruits are first heated in water for about 10 minutes to separate the segments from the stone. It also helps in removing astringency. Pulp is extracted by addition of water equal to the weight of segments and passing through pulping machine. The following two techniques of pulp extraction can be utilized as depicted in. (Singh et al. 1993).

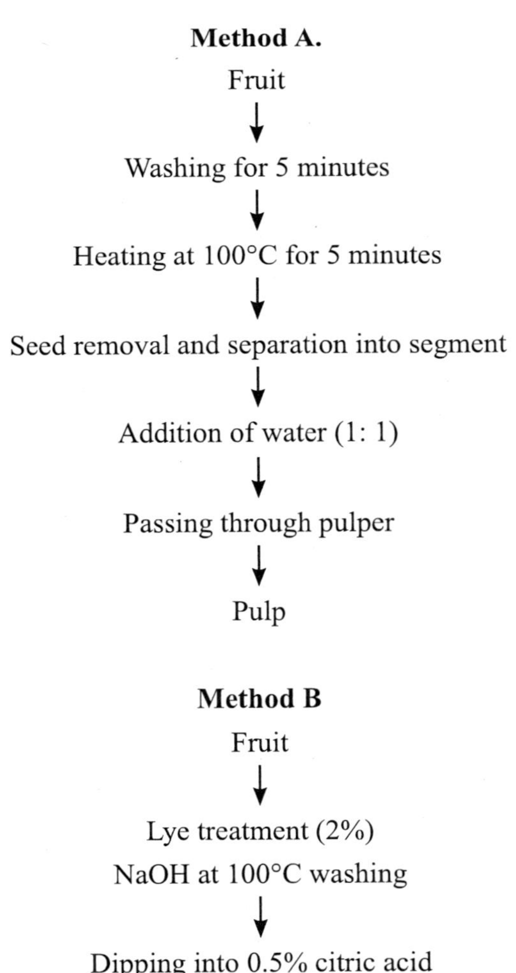

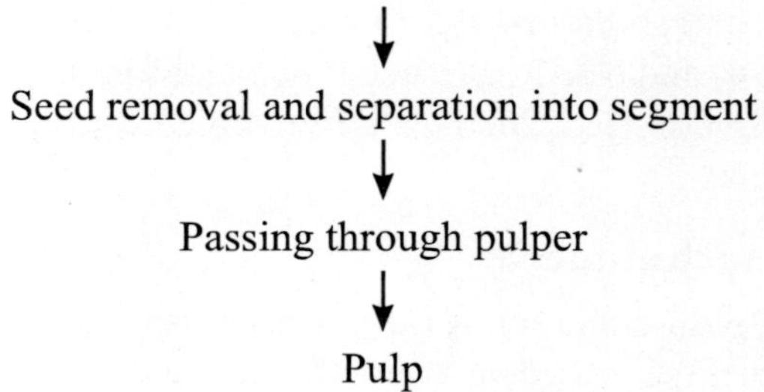

Flow Chart for Extraction of Aonla Pulp

The pulp is used for preparation of beverages, jam, and sauce. If pulp has to be preserved, it should be heated to 75°C, potassium metabisulphite (2g/kg of pulp) should be mixed thoroughly and the pulp should be filled in clean sterilized bottles or large size pouches and sealed. Enzymatic liquidification of fruits needs to be tried for extraction of clear juice. Since considerable losses in the nutritive value of aonla occur at ambient or at high temperature, it is advisable to standardize the techniques of juice extraction at low temperature under controlled conditions to minimize nutrients losses. The pomace obtained can be concentrated (pulp) and aseptically packed for further use and distant transportation.

Product Preparation

Aonla fruits can be processed into number of products. The most suitable recipe is determined by organoleptic evaluation. Some of the important products are :-

Ready to Serve (RTS)

This can be prepared by taking 10 per cent pulp and 12 per cent total soluble solids with 0.3 per cent acidity. Prepared beverage is bottled (200 ml capacity) and crown corked. Then bottles are pasteurized in boiling water for 20 minutes, air cooled and stored for use. Flow chart for preparation of ready to serve beverage is as follows:

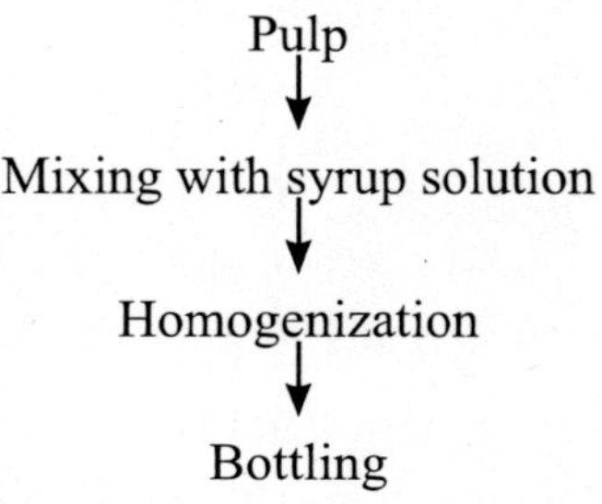

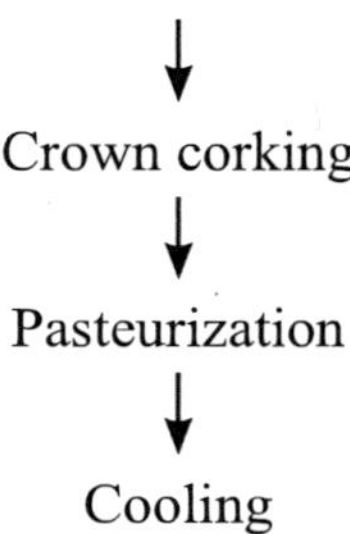

Flow Chart for Preparation of Ready to Serve (RTS) Beuerage

RTS blend can also be prepared by adding 2 percent ginger juice in pulp. Besides ginger, it can also be mixed with carrot, limes, etc. for improved nutrition and better palatability.

Squash

Squash is prepared by taking 40-45 per cent pulp and 50 per cent soluble solids with 1 per cent acidity and 350 ppm SO_2. Prepared squash is bottled in bottles of 750-ml. capacity, capped and stored for use. Flow chart for preparation of aonla squash is given in.

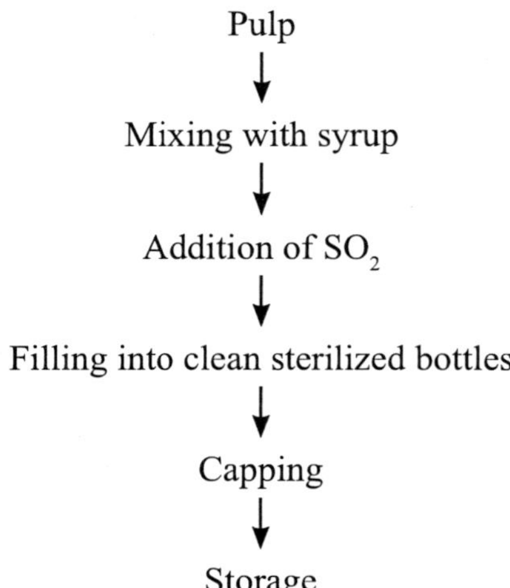

Flow Chart for Preparation of Aonla Squash

Syrup

The ideal recipe for preparation of syrup is 40-45 per cent pulp, 68 per cent total soluble solids and 1.2 per cent acidity. Prepared syrup is filled into clean and sterilized bottles (750 ml) and kept for storage. The technique for preparation of syrup is similar to that of squash.

Jam

Cultivars, which have low fibre content is most suitable for jam making. Aonla fruit jam of the composition of 45 per cent pulp with 68 per cent total soluble solids and 0.5 per cent acidity produce a highly acceptable product. Fruit pulp is mixed with sugar and citric acid and cooked to desired consistency. Prepared jam is filled into clean and sterilized bottles (500 g capacity) and these are sealed and stored for use. The technique used for preparation of aonla jam is as follows :

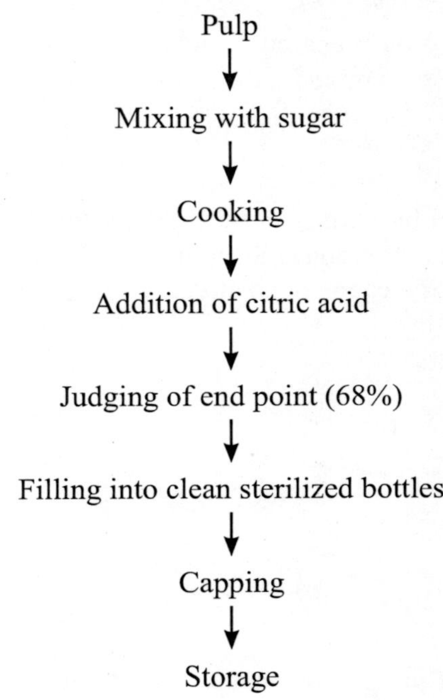

Flow Chart for Preparation of Aonla Jam

Preserve

At present, the most common use of good quality aonla fruit is making whole fruit preserve or Murabba. The major cities in India, where it is commercially prepared are Amritsar, Mumbai, New Delhi, Hathras and Pratapgarh. Lye peeling of aonla has been suggested for preserve preparation to avoid pricking and improve the quality of product (Tandon et all.2003). Aonla cultivar Krishna and NA-6 are well suited for preserve preparation. The flow chart for preparation of preserve is as follows :

Dipping in 40 per cent syrup solution for 24 hours is important. Every day fruits are taken out and the sugar syrup is boiled till it is some what thickened. The process is repeated for 3-4 times and when no more dilution of syrup takes place even after fruits are immersed in the syrup, it makes the end point. Now give the final boiling to the whole mass and cool it. Put the aonla preserve in clean dry wide mouthed jars.

Besides losses in the nutritive value, the aonla preserve, which is available in the market, are dull brown and sticky. It should be glossy bright, succulent, fibreless and in bite size pieces.

Following technique is proposed to be attempted for obtaining desirable attributes in aonla preserve: (*i*) automatic pricking machine or cell puncturing by freezing or lye peeling need to be attempted; (*ii*) modern continuous vaccum concentration method to be tried to improve colouration and glossiness; (*iii*) various glacing agent, viz., sugar, low methoxy pectin gelatin need to be tried; (*iv*) use of chelating agent, SO_2, etc. to be tried to improve colouration; and (*v*) preservation of whole aonla fruits in brine need to be standardized so that preserve can be made round the year.

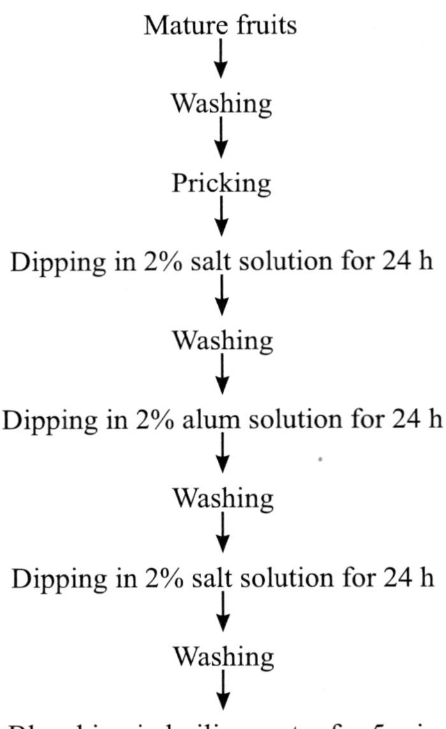

Flow Chart for Preparation of Aonla Preserve

Candy

Since, preserve is always impregnated with sticky sugar syrup, its transportation, storage and even consumption is troublesome. Hence candy has been found to be a better alternative of preserve. It will further be advisable to prepare candy of individual aonla segments. For this purpose, cultivar with low fibre content, and in which segments are easily separated are ideal one. Aonla cultivar, viz., Banarasi, Krishna and NA-6 are ideal for preparation of quality candy. Excellent quality candy can be prepared from fruits with 75 per cent total soluble solids. The flow chart for preparation of candy is depicted in Fig. below. Studies on effect of blanching and lye peeling on nutritional quality of six aonla cultivars revealed that effect of blanching was less severe than that of lye peeling on nutritional quality. A marked decrease in acidity, ascorbic acid, tannins and reducing sugars were noted due to lye peeling. Sensory scores of candy prepared from lye peeled fruits were higher due to its shining colour and taste.

Mature fruit → Washing → Pricking → Dipping in 2% salt solution for 24 hours

↓

Washing ← Dipping in 2% alum solution for 24 hours ← Weighting

↓

B1anching in boiling water for 5 minutes → Dipping in 40% syrup solution for 24 hours

↓

Dipping in 40% syrup solution for 24 hrs ← Steeping in 50% syrup solution for 24 hrs

↓

Steeping in 60% syrup solution for 24 hrs → Steeping in 70% syrup solution for 3 days

↓

Drying the fruits under shade upto 15% moisture or 75% TSS ← Draining

↓

Steeping in 60% syrup solution for 24 hours → Wrapping with powdered sugar

↓

Storage ← Packing in 400 guage polythene pouches Anola Candy

Flow Chart for Preparation of Candy

Segments in Syrup

Conventional preparation of whole aonla preserve (murabba) is a lengthy procedure, as it includes exhaustive process of pricking of fruits, requiring extra labour and cost apart from loss of nutrients. Besides, product is difficult to be eaten with the help of spoon or fork in the modern society. For segments in syrup, the segments are separated from the fruits as in case of candy. The segments are dipped overnight in 500 Brix sugar syrup containing 0.5 percent citric acid in the ratio of 1.5. The syrup is concentrated to 60° Brix by boiling and adding sugar. The segments are dipped overnight again, the syrup is reconcentrated to 70° Brix and segments are dipped overnight. Finally, segments are filled in glass/plastic jars and syrup (70° Brix) is poured in such a way that it completely covers the segments. The segments are light in colour and soft in texture and can be stored up to 12 months. The product can be eaten with the help of spoon or fork.

Pickle

Recipe containing 1 kg aonla segments, 150 g salt, 10 g each turmeric, red chillies and nigella seeds, 30 g fenugreek powder and 300 ml cooking oil considered to be the ideal for pickle.

Supari (Dried Salted Segments)

Aonla supari available in the market is usually prepared by cutting the segments of fruit into small pieces or segments, drying them under sun and then mixing with salt and other spicy ingredients, During the process, huge losses in vitarnin-C and other nutrients occur, defying the purpose of the product. Drying in open not only darkens the colour of the product, but there is possibility of contamination also from the dust and microbial spores, resulting into deterioration in the physical as well as nutritional quality of the product. The blanched separated segments of aonla are cut into small pieces (3-4 pieces/segments) and mixed thoroughly with 2.5 per cent common salt. The mixer is left overnight for osmotic dehydration of pieces. The leachet, if any, is discarded and pieces are dried in electric dehydrator at 60°C to a moisture level of 5-6 per cent. The dried supari is packed in pouches. The product could be stored up to 12 months at room temperature. Supari prepared by refined technology was light in colour and soft in texture compared to that available in the market. It retained around 50 per cent of vitamin-C present in the fresh fruit. The product can prove a good alternate to other chewing products like tobacco, pan masala, etc., which have harmful effects on human health.

Sauce

Recipe and technique for preparation of aonla sauce is similar to that of tomato sauce. Tomato pulp of high Iycopene content can be mixed with aonla pulp for better coloration and nutrition. Sauce containing 0.5 kg aonla pulp + 0.5

kg tomato pulp with 75g sugar, 10g salt, 50g onion, 5 g garlic, 10g ginger, 5 g red chillies and 10g hot spices was found to be the most ideal for preparation of sauce. Flow sheet for sauce is depicted in below. The quality of sauce remains acceptable up to 12 months at room temperature.

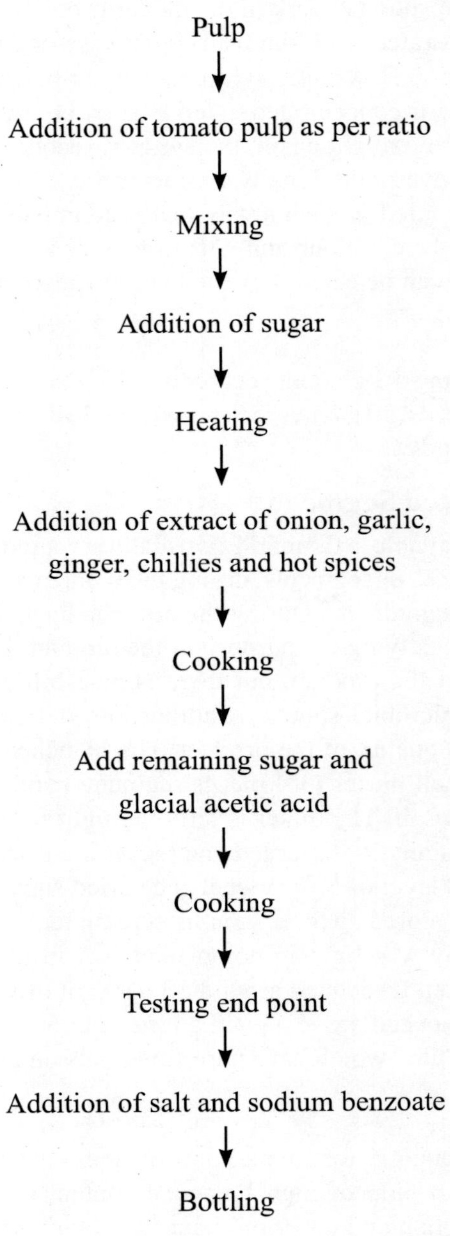

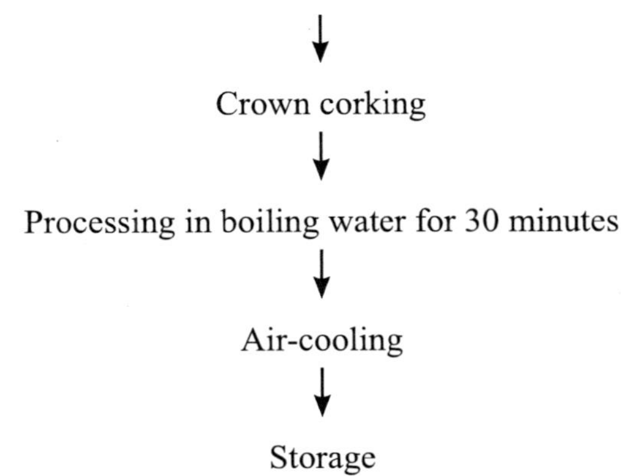

Flow Chart for Preparation of Aonla and Tomato Mixed Sauce

Aonla Churan (*Spiced Salted Powder*)

Blanched aonla segments are dried in electric dehydrator at 60°C and ground. The powder is mixed with other spices. The best recipe worked out is 8% salt, 16% black salt, 15% sugar, 3% citric acid, 2% black pepper, 1% asafoetida, 1% cumin seeds, 1% fennel seeds, 1% ginger, 0.5% azwain and 2.5% mint leaves, all in powdered form. The churan (powder) is packed in Pouches. The product can be stored for fairly long period in airtight condition.

Since churan (powder) contains 65-68 per cent aonla powder, the nutritive value of the product is high and so as the therapeutic value. Most of the ingredients have one or the other therapeutic quality. The churan could be useful to stomach upsets and could also be used in fruit salads and soups. Aonla is an accepted hair tonic, is traditional recipes for enriching hair growth and pigmentation.

Food Products

Out of all aonla food products, murabba (whole fruit preserve) is the most popular. It is recognized as a nutritious food item and is recommended for consumption during summer months. The major murabba producing units are clustered in Northern India. These units are located at Amritsar, Hathras, Varanasi, Pratapgarh and Allahabad. Instead of preserve, it appears that aonla candy in pouches and syrup, RTS, squash and jam will capture the major domestic market while freeze dried powder has immense potentiality for export in the overseas market.

Chapter 6

Custard Apple *(Annona Reticulata)*

Custard Apple is also known as sugar apple, this fruit has a lumpy green skin covering masses of sweet, scented white flesh: in most varieties the fruit can easily be divided into two pieces by hand and the creamy flesh eaten with a spoon. Both in tree and in fruit, the custard apple, *Annona reticulate L* is generally rated as the mediocre or "ugly duckling" species among the prominent members of this genus.

The custard apple tree is not especially attractive. It is erect, with a rounded or spreading crown and trunk 10 to 14 inch (25-35 cm) thick. Height ranges from 15 to 35 ft (4.5-10 m). The ill-smelling leaves are deciduous, alternate, oblong or narrow-lanceolate, 4 to 8 inch (10-20 cm) long, 3/4 to 2 inch (2 5 cm) wide, with conspicuous veins. Flowers, in drooping clusters, are fragrant, slender, with 3 outer fleshy, narrow petals 3/4 to 1 1/4 in (2-3 cm) long; light-green externally and pale-yellow with a dark-red or purple spot on the inside at the base. The flowers never fully open.

The compound fruit, 3.25 to 6.5 inch (8-16 cm) in diameter, may be symmetrically heart-shaped, lopsided, or irregular; or nearly round, or oblate, with a deep or shallow depression at the base. The skin, thin but tough, may be yellow or brownish when ripe, with a pink, reddish or brownish-red blush, and faintly, moderately, or distinctly reticulated. There is a thick, cream-white layer of custardlike, somewhat granular, flesh beneath the skin surrounding the concolorous moderately juicy segments, in many of which there is a single, hard, dark-brown or black, glossy seed, oblong, smooth, less than 0.5 inch (1.25 cm) long. A pointed, fibrous, central core, attached to the thick stem, extends more than halfway through the fruit. The flavor is sweet and agreeable though without the distinct character of the cherimoya, sugar apple, or atemoya.

Origin and Distribution

The custard apple is believed to be a native of the West Indies but it was carried in early times through Central America to southern Mexico. It has long been cultivated and naturalized as far south as Peru and Brazil. It is commonly grown in the Bahamas and occasionally in Bermuda and southern Florida.

Apparently it was introduced into tropical Africa early in the 17th century and it is grown in South Africa as a dooryard fruit tree. In India the tree is cultivated, especially around Calcutta, and runs wild in many areas. It has become fairly common on the east coast of Malaya, and more or less throughout southeast Asia and the Philippines though nowhere particularly esteemed. Eighty years ago it was reported as thoroughly naturalized in Guam. In Hawaii it is not well known.

Cultivars

No named cultivars are reported but there is considerable variation in the quality of fruit from different trees. The yellow-skinned types seem superior to the brownish, and, when well filled out, have thicker and juicier flesh. Seeds of a purple-skinned, purple-fleshed form, from Mexico, were planted in Florida and the tree has produced fruit of unremarkable quality.

Climate

The custard apple tree needs a tropical climate but with cooler winters than those of the west coast of Malaya. It flourishes in the coastal lowlands of Ecuador; is rare above 5,000 ft (1,500 m). In Guatemala, it is nearly always found below 4,000 ft (1,220 m). In India, it does well from the plains up to an elevation of 4,000 ft (1,220 m); in Ceylon, it cannot be grown above 3,000 ft (915 m). Around Luzon in the Philippines, it is common below 2,600 ft (800 m). It is too tender for California and trees introduced into Palestine succumbed to the cold. In southern Florida the leaves are shed at the first onset of cold weather and the tree is dormant all winter. Fully grown, it has survived temperatures of 27° to 28°F (-2.78° to 2.22°C) without serious harm. This species is less drought-tolerant than the sugar apple and prefers a more humid atmosphere.

Soil

The custard apple does best in low-lying, deep, rich soil with ample moisture and good drainage. It grows to full size on oolitic limestone in southern Florida and runs wild in light sand and various other types of soil in the New and Old World tropics but is doubtless less productive in the less desirable sites.

Propagation

Seed is the usual means of propagation. Nevertheless, the tree can be multiplied by inarching, or by budding or grafting onto its own seedlings or onto soursop, sugar apple or pond apple rootstocks. Experiments in Mexico, utilizing cherimoya, llama, soursop, custard apple, *Annona* sp. Af. *lutescens* and *Rollinia jimenezii* Schlecht as rootstocks showed best results when custard apple scions were side-grafted onto self-rootstock, soursop, or A. sp. Af. *lutescens*. Custard apple seedlings are frequently used as rootstocks for the soursop, sugar apple and atemoya.

Culture

The tree is fast-growing and responds well to mulching, organic fertilizers and to frequent irrigation if there is dry weather during the growing period. The form of the tree may be improved by judicious pruning.

Harvesting and Yield

The custard apple has the advantage of cropping in late winter and spring when the preferred members of the genus are not in season. It is picked when it has lost all green color and ripens without splitting so that it is readily sold in local markets. If picked green, it will not color well and will be of inferior quality. The tree is naturally a fairly heavy bearer. With adequate care, a mature tree will produce 75 to 100 lbs (34-45 kg) of fruits per year. The short twigs are shed after they have borne flowers and fruits.

Pests and Diseases

The custard apple is heavily attacked by the chalcid fly. Many if not all of the fruits on a tree may be mummified before maturity. In India, the ripening fruits must be covered with bags or nets to avoid damage from fruit bats.

A dry charcoal rot was observed on the fruits in Assam in 1947. In 1957 and 1958 it made its appearance at Saharanpur. The causal fungus was identified as *Diplodia annonae*. The infection begins at the stem end of the fruit and gradually spreads until it covers the entire fruit.

Food Uses

In India, the fruit is eaten only by the lower classes, out-of-hand. In Central America, Mexico and the West Indies, the fruit is appreciated by all. When fully ripe it is soft to the touch and the stem and attached core can be easily pulled out. The flesh may be scooped from the skin and eaten as is or served with light cream and a sprinkling of sugar. Often it is pressed through a sieve and added to milk shakes, custards or ice cream.

Table : Food Value Per 100 g of Edible Portion*

Calories	80-101
Moisture	68.3-80.1 g
Protein	1.17-2.47 g
Fat	0.5-0.6 g
Carbohydrates	20-25.2 g
Crude Fiber	0.9-6.6 g
Ash	0.5-1.llg
Calcium	17.6-27 mg
Phosphorus	14.7-32.1 mg
Iron	0.42-1.14 mg
Carotene	0.007-0.018 mg
Thiamine	0.075-0.119 mg
Riboflavin	0.086-0.175 mg
Niacin	0.528-1.190 m
Ascorbic Acid	15.0-44.4 mg
Nicotinic Acid	0.5

*Minimum and maximum levels of constituents from analyses made in Central America, Philippines and elsewhere.

Custard Apple Recipes

Custard Apple Sorbet

Ingredient

- 75 g caster sugar
- medium custard apples, halved
- Thinly pared lime rind to decorate

Method

Place the sugar in a pan with 300ml water; bring to the boil, stirring to dissolve the sugar. Reduce the heat; simmer for 10 minutes until it forms thin syrup. Set aside to cool slightly. Scoop out custard apple flesh. Press through a sieve to remove seeds. Place the sugar syrup and flesh in a food processor; blend until smooth. Pour into a freezer container and freeze for 2 hours. Remove from freezer, mash with a fork. Freeze again until firm. Serve scoops in small bowls decorated with lime rind.

Custard Apple Segments

Ingredients

- 1 large custard apple

Method

With the tip of a spoon dip in near the skin at the base of a segment, lift out and arrange the segments attractively on the sauce.

- Garnish with a carved red plum rose, and fresh leaf.
- Custard apple segments can be eaten easily with chopsticks, or with a spoon.
- Segment sizes vary Sauce: Boil a stick of cinnamon, ginger syrup with ginger cubes cut in slices

Cheesecake

Ingredients

Base

- 1 pkt ginger nut biscuits
- 125 melted butter

Filling

- 250g cream cheese
- ½ cup castor sugar
- 250 evaporated milk (cold)
- 3 teaspoons gelatin dissolved in 1/4 cup boiling water
- Juice of 1 lemon or lime
- 1 large custard apple

Method

Crush the biscuits finely and mix with the melted butter. Press into a 20cm tart tray and chill for half an hour. Dissolve the gelatin in water. Put custard apple flesh in a blender and blend until smooth. Whip the evaporated milk until thick. Mix all ingredients together. Pour into base and put in fridge to set.

Decorate with slices of kiwifruit and sprinkle shredded coconut and toasted almonds on top to serve.

Custard Apple Topped with Raspberries and Muesli

Ingredients

- 2 large custard apples (Peeled) remove seeds cut into wedges
- 1 cup frozen raspberries

- 75g (3/4cup) mixed fruit muesli
- 2 tablespoons maple syrup
- 50g butter, melted
- Vanilla yogurt to serve (optional)

Method

Preheat oven to 200C. Line a baking tray with baking paper. Lay custard apple wedges on baking tray and top with raspberries. Sprinkle muesli over, and then drizzle with maple syrup and butter. Bake for 15 minutes, or until the top is golden and fruit tender. Serve custard apple warm with vanilla yoghurt.

Custard Apple Cream

Ingredients

- 3tsps gelatin
- 1/3 cup boiling water
- 500 soft light cream cheese
- 1 cup caster sugar
- 190ml cream
- 2 custard apples, cut into segments, deseeded and pureed with a squeeze of lemon

Method

Dissolve gelatin in one-third of a cup of boiling water. In a bowl beat softened cream cheese thoroughly then add gelatin mixture, caster sugar and cream. Add custard apple puree and beat until smooth. Pour into 8 dessert glasses or pots. Refrigerate for 2-3 hours.

Custard apple Teacake

Ingredients

- 125g butter
- ½ cup + 3tsp extra, caster sugar
- 1 tsp vanilla essence
- 2 eggs
- 3 ripe custard apples, halved (about 350g each)
- 1 cups self-raising flour
- ½ tsp ground cinnamon

Method

Preheat oven to 17°C. Lightly grease a loaf pan or line with baking paper. Using electric beaters beat butter ½ cup caster sugar and vanilla in a medium bowl until sugar dissolves and mixture is pale and creamy. Add eggs one at a time. Beat well after each addition. Using a spoon remove pulp from custard apples. Discard seeds and add pulp to mixture. Mix well until combined. Using a large metal spoon, fold flour in 2 batches into mixture. Spoon into loaf pan and smooth the top.

Combine extra sugar and cinnamon and sprinkle over mixture. Bake for 1 hour or until skewer comes out clean. Remove from oven and stand for 5 minutes before turning cake out onto a wire rack to cool. Slice and serve.

Custard Apple Milk Shake

Preparation of Custard apple pulp : Good quality fully eye opened, ripened but firm fruits having uniform green colour, with creamy whitish pulp and pleasant flavor may be selected. After washing manual peeling should be done. The pulp and seeds can be separated by rubbing the pulpy seeds on stainless steel wire mesh. The pulp was filtered through muslin cloth. Then pulp is pasteurized at 85°C for 30 min and packed in high density polythene pouches. The pulp was stored at 15-20°C temperature in refrigerator for 2 hrs. and used after addition of sodium alginate.

Preparation of Custard Apple Milk shake

The diagram of preparation of custard apple milk shake is as following :

Flow Diagram for Preparation of Custard Apple Milk Shake

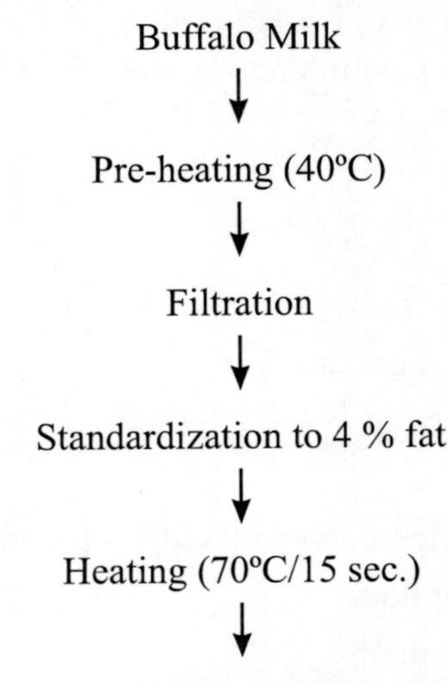

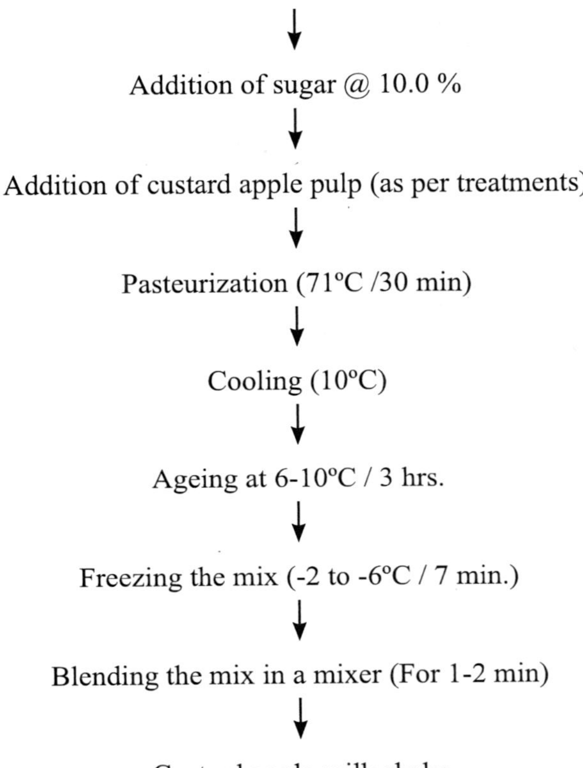
↓
Addition of sugar @ 10.0 %
↓
Addition of custard apple pulp (as per treatments)
↓
Pasteurization (71ºC /30 min)
↓
Cooling (10ºC)
↓
Ageing at 6-10ºC / 3 hrs.
↓
Freezing the mix (-2 to -6ºC / 7 min.)
↓
Blending the mix in a mixer (For 1-2 min)
↓
Custard apple milk shake

Chapter 7

Jamun *(Syzgium Cumini)*

Jambul or Jamun or Jamblang (*Syzgium cumini*) is an evergreen tropical tree in the flowering plant family Myrtaceae, native to India and Indonesia. The black plum or Indian berry, commonly known as jamun, water apple, rose apple and Malayam rose apple, are minor fruit of commercial value.

It is also grown in other areas of southern Asia including Myanmar and Afghanistan. The tree was also introduced to Florida, USA in 1911 by the USDA, and is also now commonly planted in Suriname. The various names for this fruit are (in Java) plum, jambul, jamun, jaman, black plum, faux pistachier, Indian blackberry, jambol, doowet and jambolan. Scientific synonyms include *Syzygium jambolanum* and *Eugenia cumini*.

A fairly fast growing species, it can reach heights of up to 30 m and can live more than 100 years. Its dense foliage provides shade and is grown just for its ornamental value. The wood is strong and is water resistant. Because of this it is used in railway sleepers and to install motors in wells. It is sometimes used to make cheap furniture and village dwellings though it is relatively hard to work on.

Jamun is a very common, large evergreen beautiful tree of Indian subcontinent. The scientific name of Jamun is *Eugenijambolana* or *Syzygium cumini L* and it belongs to the myrtaceae plant family. Common names are Java plum, black plum, jambul and Indian blackberry. It grows naturally in clayey loam soil in tropical as well as sub-tropical zones. It is widely cultivated in Haryana as well as the rest of the Indo-Gangetic plains on a large scale. Its habitat starts from Myanmar and extends up to Afghanistan. It is generally cultivated as a roadside avenue tree as well.

Jamun tree tends to grow as umbrella like crown having dense foliage. It thus gives pleasant cool shade during summer. It tends to have a straight bole when coming up on rich soil and favorable climate, but a crooked one when on a dry terrain and unfavorable environment. Its bark is light gray in color and fairly smooth in texture.

Jamun foliage comprises leaves measuring about 10 to 15 cm long and 4 to 6 cm wide. These are entire, ovate-oblong, sometimes lanceolate and also acuminate, coraceous, tough and smooth with shine above. The fragrant flowers of Jamun are small, nearly 5 mm in diameter. These are arranged in terminal trichotomous panicles greenish white in color. These appear during March- April.

Jamun fruit appears in May-June. The berry is oblong, ovoid, green when just appearing, pink when attaining near maturity and shining crimson black when fully ripe. The fruit of wild variety called kath-jamun or woody Jamun are small and tart in taste. The ones of grafted for improved variety are large, and deliciously sweet, but slightly sour.

The Jamun tree starts flowering in March-April. The fragrant flowers of Jamun are small, nearly 5 mm in diameter. This is followed by the fruit which appears in May-June and resembles a large berry. The berry is oblong, ovoid, green when just appearing, pink when attaining near maturity and shining crimson black when fully ripe. Another variety comes in white and is said to have medicinal properties. Jamun fruit is a mixture of sweet, slightly sub acid spicy flavour that stands out even after eaten since it turns the tongue into purple color. The fruit is universally accepted to be very good for medicinal purposes, especially diabetics. The seed is also used in various alternative healing systems like Ayurveda, Unani and Chinese medicine for digestive ailments. The leaves and bark are used for controlling blood pressure and gingivitis. Wine and vinegar are also made from the fruit. Jamun is basically not a table fruit, and hence, its commercial potential is yet to be fully tapped. Originating from Lucknow, the fruit is gradually establishing itself in certain pockets of Dindigul and in remote places in Salem.The jamun is attractive look wise, if not, taste wise. The fruit-lovers do not have any special preference for jamun because of its "sub acid spicy flavour". It comes in two varieties: seedless (yet it retains a rudimentary seed), and jumbo. It has also got another premium variety, which comes in white, which is said to have medicinal properties. Rich in iron and minerals, the fruit also acts as a coolant and induces digestive power.

Uses

Jamun tree is useful in many ways. The foliage serves as fodder, for cattle, especially during drought. The twigs form good datoon (tooth brush). The Jamun twig is also used as a rough painting brush for lettering addresses on ones gunny packs while moving household affects from one place to another. Unripe fruit is used for making vinegar. The juice is also criminating, diuretic and gives a soothing effect on human digestive system. The juice of ripe fruit is used for preparing sauces as well as beverages. It is also dried with salt and preserved as a digestive powder or churan. The seed as well as bark have several applications in Ayurveda, Unani and Chinese system of medicine. The seed is also rich in protein and carbohydrates. It also contains traces of calcium. These are, therefore, widely used as cattle feed, a medicine against diabetes and antidote in a kind of soft-food poisoning. Diabetic patients can take Jamun fruit regularly during the season of its availability for a temporary relief from the said malady. The Jamun bark also acts as tonic, astringent and anti-periodic too.

It is recommended that a mixture of equal quantity of jamun and mango juices is an ideal concoction for a diabetic. Though considered the underdog among fruits, jamun has varied utility, which is quite astounding. Mostly used for dessert purposes, jamun could be used to make varied products such as beverages, jam, jelly, squash, vinegar, and pickles.

The tasty and pleasantly flavoured jamun fruit is mostly used for dessert purpose and it is very much liked by the people. The fruit is usually shaken with salt before eating. The jamun fruit has sub acid spicy flavor. Apart from eating fresh it is also used for making delicious beverages, jellies, jam, squash, wine, vinegar and pickles. Jamun squash is very refreshing drink in the summer season. A little quantity of fruit syrup is useful for curing diarrhoea. A mixture of jamun juice and mango juice in equal quantity is very good for quenching thirst for diabetic patient. Jamun fruit is used for preparation of wine. The vinegar prepared from juice extracted from slightly unripe fruit is stoma chic, carminative and diuretic, apart from having cooling and digestive properties.

Table 7.1 : Composition of Jamun Fruit (per 100g)

Composition	*Value (g)*
Moisture	83.20
Reducing sugar	14.00
Crude fibre	0.90
Ascorbic acid	0.25
Anthocyanin	0.14
Ash	0.33
Nitrogen	0.13
Tannin	1.90
Fat	0.30

Flowchart for Processing of Squash

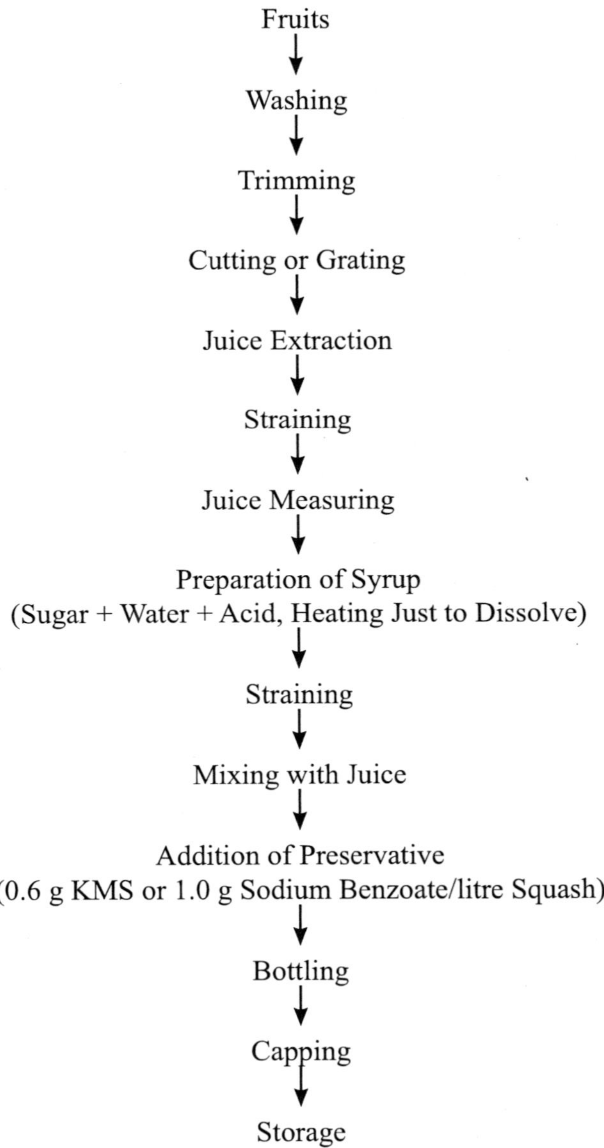

Ready To Serve Beverage

This is a type of fruit beverage which contains at least 10 per cent fruit juice and 10 per cent total soluble solids besides about 0.3 per cent acid. It is not diluted before serving, hence it is known as ready-to-serve (RTS).

Flow-Sheet for Processing of RTS Beverages

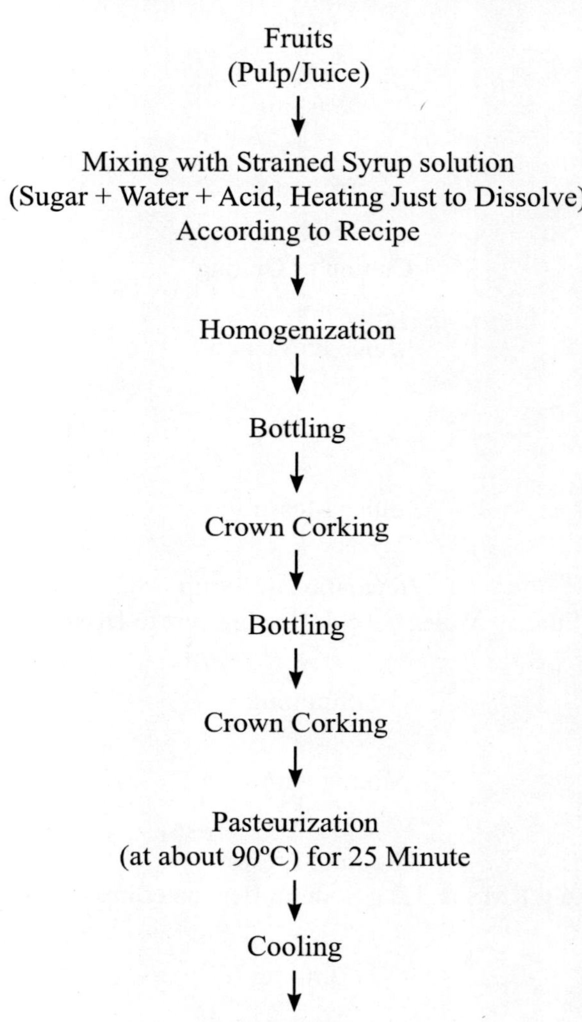

C. Cordial

It is a sparkling, clear, sweetened fruit juice from which pulp and other insoluble substances have been completely removed. It contains at least 25 per cent juice and 30 per cent TSS. It also contains about 1.5 per cent acid and 350 ppm of sulphur dioxide. This is very suitable for blending with wines. Lime and lemon are suitable for making cordial.

Flowchart for Processing of Cordial

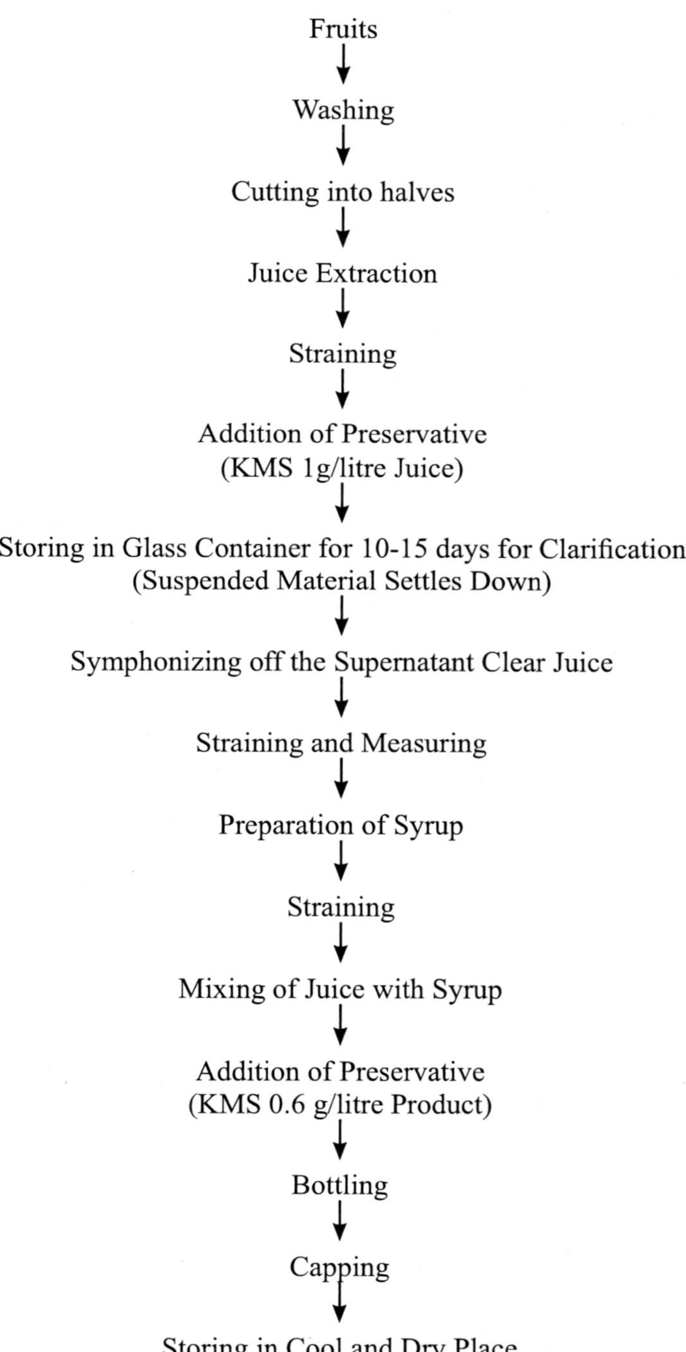

Flowchart for Making Jamun Wine

Jamun Fruits

↓

Washing and cleaning by tap water and dip in 5% NaCl salt solution for 72 hour

↓

Separate the seed from the fruit

↓

Crushed the fruit pulp with water (1:1 ratio) in a Mixture-cum- Grinder

↓

Pressing and extraction of the juice; Add SMS (100 µg.ml-1
Must amelioration

↓

Adjust TSS to 17 ^{0} Brix with cane sugar. Acidify the must to pH 4.5 using 1N acetic acid and Inoculation with starter culture

↓

Inoculate with wine yeast starter culture (use 28-48 hour old starter culture at 2% (v/v)

↓

Fermentation (at 32±2°C for 6 days)

↓

Racking and decantation. Carrying out first racking when the Brix reaches 2-3°. Two- three more racking at 15 days intervals if sedimentation persists

↓

Clarification (add 0.04% bentonite)

↓

Final racking

↓

Bottling and corking. Fill in bottle full, Cork and seal the bottle with bees wax

↓

Wine

Chapter 8

Guava *(Psidium Guajava L)*

One of the most gregarious of fruit trees, the guava, *Psidium guajava* L., of the myrtle family (Myrtaceae), is almost universally known by its common English name or its equivalent in other languages. In Spanish, the tree is *guayabo,* or *guayavo,* the fruit *guayaba* or *guyava* The French call it *goyave* or *goyavier;* the Dutch, *guyaba, goeajaaba;* the Surinamese, *guave* or *goejaba;* and the Portuguese, *goiaba* or *goaibeira.* Hawaiians call it guava or *kuawa.* In Guam it is *abas.* In Malaya, it is generally known either as guava or *jambu batu,* but has also numerous dialectal names as it does in India, tropical Africa and the Philippines where the corruption, *bayabas,* is often applied. Various tribal *names-pichi, posh, enandi, etc,-are* employed among the Indians of Mexico and Central and South America.

Guava (*Psidium guajava*), Fresh, Nutritive Value per 100g.

Principle	*Nutrient Value*	*Percentage of RDA*
Energy	68 Kcal	3.5%
Carbohydrates	14.3 g	11.5%
Protein	2.55 g	5%
Total Fat	0.95 g	3%
Cholesterol	0 mg	0%
Dietary Fiber	5.4 g	14%
Vitamins		
Folates	49 µg	12.5%
Niacin	1.084 mg	7%
Pantothenic acid	0.451 mg	9%
Pyridoxine	0.110 mg	8.5%

Principle	*Nutrient Value*	*Percentage of RDA*
Riboflavin	0.040 mg	3%
Thiamin	0.067 mg	5.5%
Vitamin A	624 IU	21%
Vitamin C	228 mg	396%
Vitamin E	0.73 mg	5%
Vitamin K	2.6 µg	2%
Electrolytes		
Sodium	2 mg	0%
Potassium	417 mg	9%
Minerals		
Calcium	18 mg	2%
Copper	0.230 mg	2.5%
Iron	0.26 mg	3%
Magnesium	22 mg	5.5%
Manganese	0.150 mg	6.5%
Phosphorus	11 mg	2%
Selenium	0.6 mcg	1%
Zinc	0.23 mg	2%
Phyto-nutrients		
Carotene-ß	374 µg	--
Crypto-xanthin-ß	0 µg	--
Lycopene	5204 µg	--

Ascorbic acid mainly in the skin, secondly in the firm flesh, and little in the central pulp-varies from 56 to 600 mg. It may range up to 350-450 mg in nearly ripe fruit. When specimens of the same lot of fruits are fully ripe and soft, it may decline to 50-100 mg. Canning or other heat processing destroys about 50% of the ascorbic acid. Guava powder containing 2,500-3,000 mg ascorbic acid was commonly added to military rations in World War II. Guava seeds contain 14% of aromatic oil, 15% protein and 13% starch. The strong odor of the fruit is attributed to carbonyl compounds.

Guavas are outstanding in their high vitamin C content, which in some varieties can be as high as five times that of fresh orange juice. The fruit freezes exceptionally well and lends itself admirably for processing. The greatest commercial uses of the fruit are in jams, jellies, guava paste and canned products.

Guava (*Psidiumguajava* L.) is a member of the large Myrtaceae or Myrtle family, believed to be originated in Central America and the southern part of Mexico (Somogyiet al. 1996). It is claimed to be the fourth most important fruit in terms of area and production after mango, banana and citrus. India is the major world producer of guava. It has been in cultivation in India since early 17th century and gradually becamea crop of commercial importance. Guava is quite hardy, prolific bearer and highly remunerative even without much care. It is widely grown all over the tropics and sub-tropics including India viz., Uttar Pradesh, Bihar, Madhya Pradesh, Maharashtra, Andhra Pradesh, Tamil Nadu, West Bengal, Assam, Orissa, Kamartaka, Kerala, Rajasthan and many more states. Main Varities grown in India are *Allahabad Safeda, Lucknow* 49, *Chittidar, Nagapur Seedless, Bangalore, Dharwar, AkraMridula, ArkaAmulya, Harijha, Hafshi, Allahabad Surkha CISHG1, CISHG2, CISHG3* etc (NHB, 2010).

Guava is often marketed as "super-fruits" which has a considerable nutritional importance in terms of vitamins A and C with seeds that are rich in omega-3, omega-6 polyunsaturated fatty acids and especially dietary fiber, riboflavin, as well as in proteins, and mineral salts. The high content of vitamin C (ascorbic acid) in guava makes it a powerhouse in combating free radicals and oxidation that are key enemies that cause many degenerative diseases. The anti-oxidant virtue in guavas is believed to help reduce the risk of cancers of the stomach, esophagus, larynx, oral cavity and pancreas. The vitamin C in guava makes absorption of vitamin E much more effective in reducing the oxidation of the LDL cholesterol and increasing the (good) HDL cholesterol. The fibers in guavas promote digestion and ease bowel movements. The high content of vitamin A in guava plays an important role in maintaining the quality and health of eyesight, skin, teeth, bones and the mucus membranes.

Food Uses

Raw guavas are eaten out-of-hand, but are preferred seeded and served sliced as dessert or in salads. More commonly, the fruit is cooked and cooking eliminates the strong odor. A standard dessert throughout Latin America and the Spanish-speaking islands of the West Indies is stewed guava shells (cascos de guayaba), that is, guava halves with the central seed pulp removed, strained and added to the shells while cooking to enrich the syrup. The canned product is widely sold and the shells can also be quick-frozen. They are often served with cream cheese. Sometimes guavas are canned whole or cut in half without seed removal. Bars of thick, rich guava paste and guava cheese are staple sweets, and guava jelly is almost universally marketed. Guava juice, made by boiling sliced, unseeded guavas and straining, is much used in Hawaii in punch and ice cream sodas. A clear guava juice with all the ascorbic acid and other properties undamaged by excessive heat, is made in South Africa by trimming and mincing guavas, mixing with a natural fungal enzyme (now available under various trade names), letting stand for 18 hours at 120° to 130° F (49°-54° C) and filtering. It is made into syrup for use on waffles, ice cream, puddings and in milkshakes. Guava juice and nectar are among the numerous popular canned or bottled fruit beverages of the Caribbean area. After washing and trimming of the floral remnants, whole guavas in syrup or merely sprinkled with sugar can be put into plastic bags and quick-frozen.

There are innumerable recipes for utilizing guavas in pies, cakes, puddings, sauce, ice cream, jam, butter, marmalade, chutney, relish, catsup, and other products. In India, discoloration in canned guavas has been overcome by adding 0.06% citric acid and 0.125% ascorbic acid to the syrup.

Dehydrated guavas may be reduced to a powder which can be used to flavor ice cream, confections and fruit juices, or boiled with sugar to make jelly, or utilized as pectin to make jelly of low-pectin fruits. India finds it practical to dehydrate guavas during the seasonal glut for jelly- manufacture in the off-season.

Green mature guavas can be utilized as a source of pectin, yielding somewhat more and higher quality pectin than ripe fruits.

Flowchart for Making Guava Products

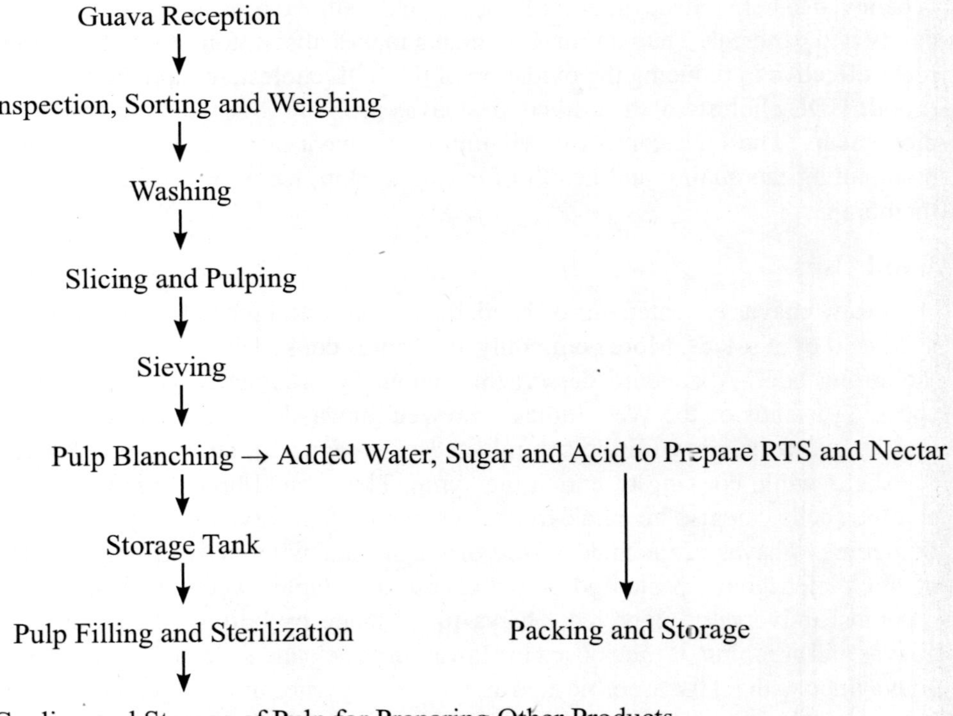

Flow-Sheet Extraction of Guava Pulp

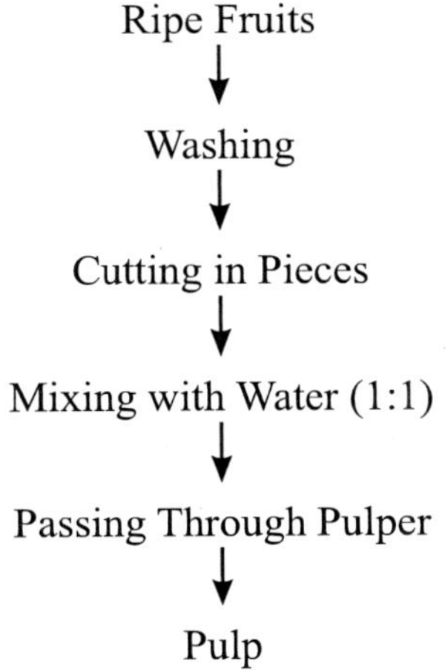

Flow-Sheet for Processing of RTS Beverages

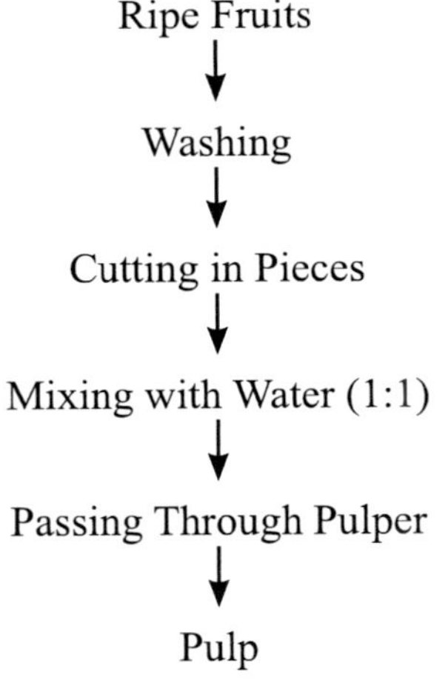

Guava Jelly

Ingredients

Guava	1 kg
Sugar	600 g
Citric acid	8 g

Method

1. Select well matured sound guava fruits, wash well and cut into small slices.
2. To this add equal quantities of water and boil under low flame for 30 minutes.
3. Strain the pectin extracts, add sugar and citric acid.
4. Boil the pectin extract and sugar mixture upto 65°Brix and fill hot into sterilized bottles, cool and store.

Guava-Lime-Ginger RTS

Flow Chart for the Preparation of Guava-Lime-Ginger RTS

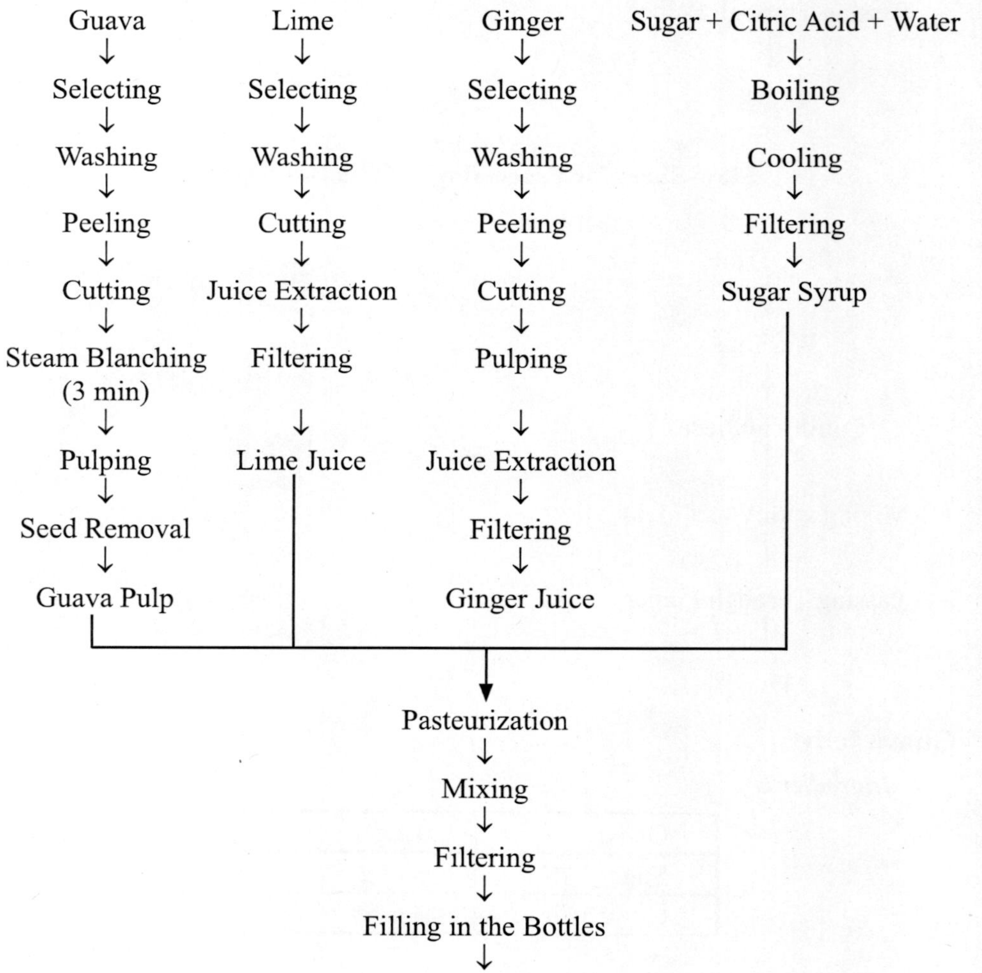

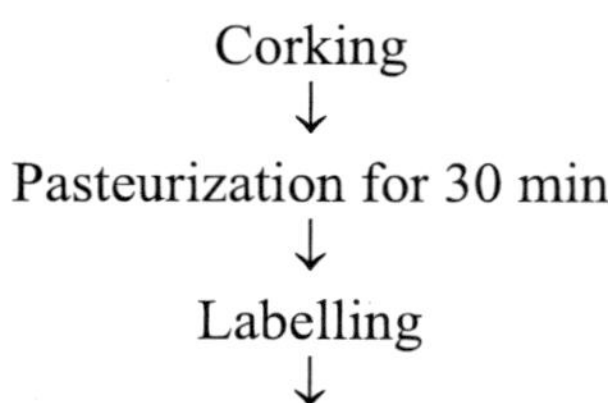

Corking
↓
Pasteurization for 30 min
↓
Labelling
↓
Storing at Room and Refrigeration Temperature

Guava-Banana RTS

Flow Chart for the Preparation of Guava-Banana RTS

Guava
↓
Selecting
↓
Washing
↓
Peeling
↓
Cutting
↓
Steam Blanching (3 min)
↓
Pulping
↓
Seed Removal
↓
Guava Pulp

Banana
↓
Selecting
↓
Peeling
↓
Cutting
↓
Pulping
↓
Banana Pulp

Sugar + Citric Acid + Water
↓
Boiling
↓
Cooling
↓
Filtering
↓
Sugar Syrup

Guava Pulp + Banana Pulp + Sugar Syrup
↓
Pasteurization
↓
Mixing
↓
Filtering
↓
Filling in the Bottles
↓
Corking
↓

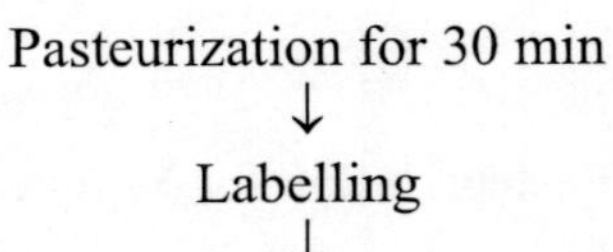

Storing at Room and Refrigeration Temperature

Mixed Fruit Squash

Flow Chart for the Preparation of Mixed Fruit Squash (Guava, Banana, Mango)

Guava	Banana	Mango	Sugar + Citric Acid + Water
↓	↓	↓	↓
Selecting	Selecting	Selecting	Boiling
↓	↓	↓	↓
Washing	Peeling	Washing	Cooling
↓	↓	↓	↓
Peeling	Cutting	Peeling	Filtering
↓	↓	↓	↓
Cutting	Pulping	Cutting	Sugar Syrup
↓	↓	↓	
Steam Blanching (3 min)	Banana Pulp	Pulping	
↓		↓	
Pulping		Mango Pulp	
↓			
Seed Removal			
↓			
Guava Pulp			

↓

Mixing
↓
Filtering
↓
Filling in the Bottles
↓
Capping
↓
Labelling
↓
Storing at Room and Refrigeration Temperature

Dehydration of Guava Pieces

Selection of Guava
↓
Washing
↓
Cutting into Eight Portions Vertically
↓
Scooping Out the Seeds
↓
Drying in Cabinet Drier (60°C)
↓
Cooling to Room Temperature
↓
Tempering Overnight
↓
Packing and Storage

Chapter 9

Woodapple (*Feronia Limonia*)

The Woodapple, *Feronia limonia* Swingle (syns. *F. elephantum* Correa; *Limonia acidissima L: Schinus limonia* L.) is the only species of its genus, in the family Rutaceae. Besides wood-apple, it may be called elephant apple, monkey fruit, curd fruit, *kath bel* and other dialectal names in India. In Malaya it is *gelinggai* or *belinggai;* in Thailand, *ma-khwit;* in Cambodia, *kramsang;* in Laos, *ma-fit.* In French, it is *pomme d' elephant, pomme de bois,* or *citron.*

Limonia acidissima is a large tree growing to 9 metres (30 ft) tall, with rough, spiny bark. The leaves are pinnate, with 5-7 leaflets, each leaflet 25–35 mm long and 10–20 mm broad, with a citrus-scent when crushed. The fruit is a berry 5–9 cm diameter, and may be sweet or sour. It has a very hard rind which can be difficult to crack open, and contains sticky brown pulp and small white seeds. The fruit looks similar in appearance to fruit of Bael (Aegle marmelos).

Nutritive Values of Wood Apple

	Pulp (ripe)	*Seeds*
Moisture	74.0%	4.0%
Protein	8.00%	26.18%
Fat	1.45%	27%
Carbohydrates	7.45%	35.49%
Ash	5.0%	5.03%
Calcium	0.17%	1.58%
Phosphorus	0.08%	1.43%
Iron	0.07%	0.03%
Tannins	1.03%	0.08%

A hundred gm of fruit pulp contains 31 gm of carbohydrate and two gm of protein, which adds up to nearly 140 calories. The ripe fruit is rich in beta-carotene, a precursor of Vitamin A; it also contains significant quantities of the B vitamins thiamine and riboflavin, and small amounts of Vitamin C.

The pulp represents 36% of the whole fruit. The pectin content of the pulp is 3 to 5% (16% yield on dry-weight basis). The seeds contain bland, non-bitter, oil high in unsaturated fatty acids.

Food Uses

The fruit is eaten plain, blended into an assortment of drinks and sweets, or well-preserved as jam. The scooped-out pulp from its fruits is eaten uncooked with or without sugar, or is combined with coconut milk and palm-sugar syrup and drunk as a beverage, or frozen as an ice cream. It is also used in *chutneys* and for making Fruit preserves jelly and jam.

Indonesians beat the pulp of the ripe fruit with palm sugar and eat the mixture at breakfast. The sugared pulp is a foundation of sherbet in the subcontinent. Jam, pickle, marmalade, syrup, jelly, squash and toffee are some of the foods of this multipurpose fruit. Young bael leaves are a salad green in Thailand. Indians eat the pulp of the ripe fruit with sugar or jaggery. The ripe pulp is also used to make chutney. The raw pulp is varied with yoghurt and make into raita. The raw pulp is bitter in taste, while the ripe pulp would be having a smell and taste that's a mixture of sourness and sweet.

The rind must be cracked with a hammer. The scooped-out pulp, though sticky, is eaten raw with or without sugar, or is blended with coconut milk and palm-sugar syrup and drunk as a beverage, or frozen as an ice cream. It is also used in chutneys and for making jelly and jam. The jelly is purple and much like that made from black currants.

Nectar from the Wood Apple Pulp

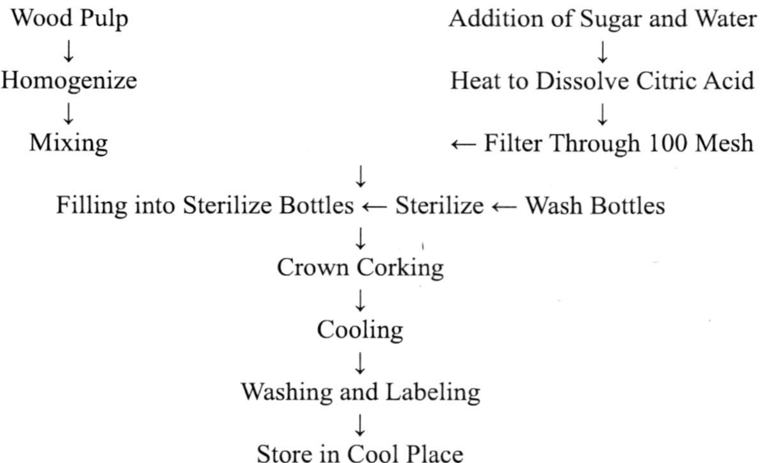

Bottled nectar is made by diluting the pulp with water, passing through pulper to remove seeds and fiber, further diluting, straining, and pasteurizing. A clear juice for blending with other fruit juices can be obtained by clarifying the nectar with Pectin. Pulp sweetened with syrup of cane or palm sugar, has been canned and sterilized. The pulp can be freeze-dried for future use but it has not bee unsatisfactorily dried by other methods.

Juice from Wood Apple

Ripe Fruit

↓

Washed with Clean Water

↓

Break the Fruit

↓

Scooping the Fruit with Seeds and Fiber (Discard Peel)

↓

Add Water

↓

Boiling

↓

Cooling

↓

Pass Through a Muslin Cloth

↓

Extracted Wood Pulp

↓

Addition of Sugar

↓

Boiling (30 min)

↓

Juice

↓

Fill into the Sterilize Cans

↓

Cooling

↓

Stored at Room Temperature

Wood Apple Chutney

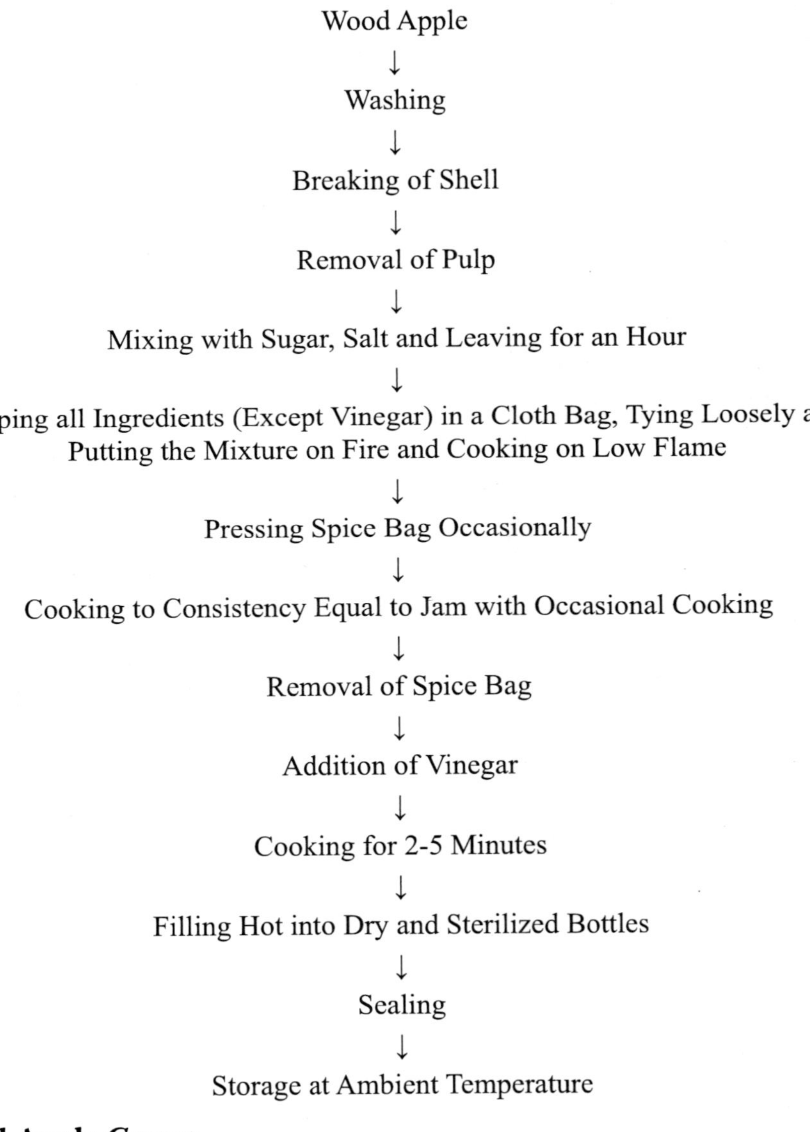

Wood Apple Cream

Wood apple cream is the scooped-out pulp of the wood-apple, though sticky, is eaten raw with or without sugar, or is blended with coconut milk and palm-sugar syrup and drunk as a beverage, or frozen as an ice cream. It is also used in chutneys and for making jelly and jam. The jelly is purple and much like that made from black currants.

Chapter 10

Bael (*Aegle marmelos*)

Bael (*Aegle marmelos* Comea) is an important indigenous fruit of India. Although the fruit is called as poor man's fruit, it is nutritive and delicious. The fruit is rich in ascorbic acid, sugar, protein, vitamin B-complex and minerals. It is reported to contain 8-100mg of ascorbic acid/100g of pulp, sugar 3-7%, TSS 12-18° Brix protein 0.8-1.5% and starch 70-80%.

There is mention of bael fruit in Vedas, Ramayana and also in Buddhist and Jain literatures. In India, It is found growing as wild in Uttar Pradesh, Madhya Pradesh, Bihar, Orissa and West Bengal. There is no systematic plantation of bael except in Uttar Pradesh. The main cultivated varieties are Mirzapuri, Kagzi, Gonda, Kagzi Banarasi, Kagzi Etawah, Narendra Bael-1 and Narendera Bael-2. All parts of this fruits three viz root, bark, leaves, flowers or fruits are used for curing one or the another human aliment. Marmelosin, a potent drug, is probably the most therapeutically active principle of bael fruit.

The bael tree is one of the most useful medicinal plants of India Its medicinal properties have been described in the *ancient medical* treatise in Sanskrit, *Charaka Samhita. All* the parts of this tree *including stem,* bark, root, leaves and fruit at all stages of maturity has medicinal virtues and has been used as traditional medicine for a long time.

The fruit is of considerable medicinal value when it just begins to ripen The ripe fruit is aromatic, astringent which helps *construction of* skin, coolant and laxative The unripe or half-ripe fruit is astringent, digestive stomachic which improves appetite and antiscorbutic, i.e. which helps to fight scurvy caused due to vitamin C deficiency.

Food Value

Bael fruit, because of its hard shell, mucilaginous texture and numerous seeds in pulp is difficult to eat out of hand and is not popular as a dessert fruit. Because of it rich aroma, which is not destroyed even during processing, bael, has a great-untapped potential for making several value added products. Further, bael fruit flavor is entirely unknown in internal and export markets. So, value added products from bael can attract both internal and export markets because there is always a demand from the consumers all over the world for new food products of its therapeutic, nutrition and delicately flavourous values. If the utilization of bael fruit is established in the processed industry, it will give a great fillip to the commercial cultivation of this valuable fruit. Hence, scientific knowledge about processing technology of bael should be disseminated on a countrywide scale.

Food Value (Per 100gm)		***Minerals and Vitamin***	
Moisture	61.5%	Calcium	85 mg
Protein	1.8%	Phosphorus	50 mg
Fat	0.3%	Iron	0.6 mg
Minerals	1.7%	Vitamin C Small amounts of vitamin B complex	8 mg
Fibre	2.9%		
Carbohydrates	31.8%		
	100%	**Calorific value - 137**	

An analysis of the bael fruit shows it is rich in mineral and vitamin contents. The sherbet made out of this fruit has all the important nutrients and health growing ingredients. It should be thick and syrupy enough to be taken with spoon and it should be thoroughly masticated. If taken hurriedly, it may produce heaviness in the stomach. The bael fruit should also not be taken in excess at a time as excessive intake of bael may produce a sensation of heaviness in the stomach and may cause gastric discomfort.

The pulp also contains a balsam-like substance, and 2 furocoumarins-psoralen and marmelosin ($C_{13}H_{12}O_3$), highest in the pulp of the large, cultivated forms. There is as much as 9% tannin in the pulp of wild fruits, less in the cultivated types. The rind contains up to 20%. Tannin is also present in the leaves, as is skimmianine. The essential oil of the leaves contains d-limonene, 56% α-d-phellandrene, cineol, citronellal, citral; 17% p-cymene, 5% cumin aldehyde.

Uses

Bael is a very good source of protein, which is 5.12 per cent of the edible portion. Fresh half-ripe bael fruit is mildly astringent and is used for dysentery and diarrhea. The pulp may be eaten or the decoction administered. Bael is said to cure without creating any tendency to constipation. Bael leaves, fruits and root can be used as tonic and coolant with antibiotic properties.

Bael fruits are valuable for its rich nutritive, sweet, aromatic mucilage and pectin contents - very good for all kinds of stomach disorders. Bael Fruits are very useful in chronic diarrhea and dysentery, particularly in the case of patients having diarrhea, alternating with the spells of constipation. Sweet drink (sherbet) prepared from the pulp of the Bael fruits produce a soothing effect on the patients who have just recovered from bacillary dysentery.

The pulp from unripe Bael fruits are soaked in gingelly oil for a week and this oil is smeared over the body before bathing. The unripe and half-ripe fruits improve appetite and digestion (Jain, 1968; Jauhari, 1969). As per Indian Ayurvedic concept this oil is said to be useful in removing the peculiar burning sensation in the soles. Rind is used for acute and amoebic dysentery, griping pain in the loins and constipation, gas, and colic, sprue, scurvy.

Peoples in South India use the juice of bael leaves to get relief from wheezing and respiratory spasm. The leaf juice is mixed in warm water with a little pepper and given as a drink.

Various parts of the tree are used for its curative, pesticidal and nutritive properties. Fresh half ripe Bael fruit is mildly astringent and used to cure dysentery, diarrhea, hepatitis, tuberculosis, dyspepsia and good for heart and brain. Roots have antidiarrhoetic, antidote to snake venom, anti- inflammatory and wound healing properties. The Bael fruit is one of the most nutritious fruits, rich in riboflavin and used for the preparation of a number of products like candy, squash, toffee, slab, pulp powder and nectar. The leaves and seed oil have pesticidal properties.

Food Uses

Bael fruits may be cut in half, or the soft types broken open, and the pulp, dressed with palm sugar, eaten for breakfast, as is a common practice in Indonesia. The pulp is often processed as nectar or "squash" (diluted nectar). A popular drink (called "sherbet" in India) is made by beating the seeded pulp together with milk and sugar. A beverage is also made by combining bael fruit pulp with that of tamarind. These drinks are consumed perhaps less as food or refreshment than for their medicinal effects.

Mature but still unripe fruits are made into jam, with the addition of citric acid. The pulp is also converted into marmalade or syrup, likewise for food and therapeutic use, the marmalade being eaten at breakfast by those convalescing from diarrhea and dysentery. A firm jelly is made from the pulp alone, or, better still, combined with guava to modify the astringent flavor. The pulp is also pickled.

Bael pulp is steeped in water, strained, preserved with 350 ppm $S0_2$, blended with 30% sugar, then dehydrated for 15 hrs at 120^0 F (48.89^0 C) and pulveri zed. The powder is enriched with 66 mg per 100 g ascorbic acid and can be stored for 3 months for use in making cold drinks (squashes). A confection, bael fruit toffee, is prepared by combining the pulp with sugar, glucose, skim milk powder and hydrogenated fat. Indian food technologists view the prospects for expanded bael fruit processing as highly promising.

The young leaves and shoots are eaten as a vegetable in Thailand and used to season food in Indonesia. They are said to reduce the appetite. An infusion of the flowers is a cooling drink.

Preparation of Products

Bael fruits can be processed into several value added products viz preserve (murabba), candy, dehydrated bael, bael powder, panjiri, jam, slab/leather and toffee. Among these products, preserve, candy, dehydrated bael, bael powder and panjiri are prepared from mature bael fruits during December - February, whereas, jam, slab and toffee can be prepared .

Preserve of Murabba

Preserve is prepared front mature, whole or large pieces of fruits in which sugar is impregnated till it becomes tender and transparent. Minimum fruits portion and minimum total soluble solids in preserve should be 55 and 70 percent, respectively.

Recipe

- Mature bael fruits slices 1 kg
- Sugar 1.25 kg
- Citric acid 2-3g
- Water 1 L

Method

Select mature, healthy bael fruits wash and cut 2 cm thick slices of fruits pulp after removing hard shell of fruits. Prick these thick slices of pulp, with stainless steel fork from both the sides and remove seeds and gummy material also from these slices. Dip pricked slices in 2 percent lime solution (20gm lime powder per litre of water) for 2-3 hours. Remove fruit slice from lime solution and wash thoroughly with clear water put the slices in a muslin cloth and dip in boiling water till they become soft. Prepare 40 percent sugar syrup using 700g sugar, 1 litre water and

2 to 3 g citric acid and dip softened fruits slices in sugar syrup for one day. Next day, drain the sugar syrup, add half of the remaining sugar, concentrate and again add in fruits slices. After two days drain syrup, add rest of sugar, concentrate syrup upto 70 percent TSS and mix in fruit slices. Maintain this concentration of sugar syrup in preserve for one week and concentrate sugar syrup, if necessary. Fill the prepared bael preserve in well sterilized wide mouth glass jars, seal with cap and store in a cool dry place.

Candy

A fruit of its pieces impregnated with sugar of glucose syrup, subsequently drained free of syrup and dried is known as candied fruit. The total sugar content the impregnated fruit is kept at about 75 percent to prevent fermentation. A certain in amount (25 to 30 percent) of invert sugar of glucose may be substituted for cane sugar to prevent crystallization of sugar. The fruit slices are drained subsequently free of syrup and dried at 55 to 60°C for 8 to 10 hours in oven. The candied bael fruit slices are then packed in polyethylene bags or glass jars and stored in cool and dry place for use in future.

Dehydrated Bael

Select mature green bael fruits wash and cut 1 to 1.5 cm thick slices of fruits pulp after removing its hard shell. Fumigate these slices of fruits with sulphur dioxide fumes for an hour in sulphur box and dehydrate at 55 to 60°C in oven upto a constant weight. Pack dried slices in polyethylene bags or glass jars for future use.

Panjiri

Bael panjiri is highly nutritive restorative and is prescribed for stomach ailments. It is prepared from bael powder, desi ghee, sugar powder, roasted wheat flour and dry fruits.

Recipe

- Bael Powder 1 kg
- Desi ghee 1 kg
- Sugar powder 1.5 kg
- Roasted wheat flour
- Dry fruits

Method

Roast bael powder in desi ghee as per recipe. Roast wheat flour as per taste in small amount of desi ghee. Mix thoroughly all ingredients viz. roasted bael powder, roasted wheat flour, sugar powder and dry fruits. Store in stainless steel utensils after making laddoos of the prepared product.

Bael Powder

Bael powder can be prepared by simply grinding dried fruit slices in a grinder. Pack ground bael powder in polyethylene bags and store in dry places after proper sealing for consumption in future.

Jam

Jam is concentrated fruit product possessing a fairly heavy body, rich in natural fruit flavour. Pectin in fruit gives it a good set and high concentration of sugar facilitaties its preservation. It is prepared by boiling the fruit pulp and a reasonably thick consistency to hold the fruit tissues in position. A fruit jam should contain minimum 45 percent of fruit portion and minimum 68 percent of total soluble solids.

Recipe

- Bael Pulp / Juice 1 kg
- Sugar 750 gm
- Citric acid 3-4 gm

Method

Select fully ripe bael fruits, wash and break fruits by striking against any hard surface. Collect fruit pulp from broken fruits and remove hard shell. Add 300-400 ml of water for each 1 kg of fruit pulp and heat up to 80°C for 1 minute after mixing thoroughly. Strain fruit pulp through stainless steel sieve arid collect pulp free from seeds and fibers. Weight fruit pulp and mix sugar and citric acid as per recipe. Cook with continuous stirring with a spoon until end point is reached. End point can be judged by doing sheet test of boiling mass. For sheet test, take boiling mass on a ladle, cool it and allow it to fall. If cooled mass falls in sheet form, end point is reached. If sheet does not form, further cooking after end point is reached and fill hot jam into clear, well sterilized, jam bottles and seal immediately. Invert jam filled bottles for 10 minutes to sterilize cap of jam bottle, label and store them in cool and dry place.

Slab

It is also known as leather or paper. Ripe bael fruits are used in its preparation. Wash ripe fruits and collect fruit pulp by breaking fruits and removing its hard shell. Add 200 to 300 ml of water for each 1 kg of fruit pulp, mix well and heat it up to 80°C. Collect fruit pulp free of seeds and fibers by straining heated mass through stainless steel sieve. Add sugar, citric acid and potassium metabisulphite (KMS) to this pulp so that treated pulp contains 35 percent total soluble solids, 0.5 percent total acidity and 0.07 percent KMS. Oil treated pulp and spread on aluminum trays smeared with butter. Dry at 55 to 60°C for 15-16 hours to moisture content of 14.5 percent. Cut slabs of dried pulp in aluminum trays, wrap in butter paper and pack in polythene bags.

Toffee

Bael toffee is prepared by using fruit pulp, sugar, glucose, skim milk powder and butter. These are very tasty having rich aroma of bael fruits.

Receipe

- Bael pulp 1 kg
- Sugar 500 gm
- Glucose 100 gm
- Skim milk powder 150 gm
- Butter 100 gm

Method

Wash ripe bael fruits and collect fruit pulp after breaking fruits. Add 750 ml water for each 1 kg of fruits pulp, mix well and heat up to 80°C. Screen homogenized pulp through stainless steel sieve and collect fine pulp free of seeds and fibers. Weigh fruit pulp as per recipe and cook it until it remains one third of its original volume. Add sugar, glucose and butter to cooking mass and again cook for some time. Now add skim milk powder to cooking mass after mixing it thoroughly in small cooking mass starts leaving sides of pan like halwa. Stop cooking and spread cooked mass uniformly in 0.50 to 0.75 cm thick layer on aluminum trays smeared with butter and allow it to cool. Cut into pieces of suitable size, wrap in moisture proof or butter paper, pack in polyethylene bags or glass jars and store at dry place.

Chapter 11

Mango (*Mangifera spp*)

The mango *(Mangifera* spp.; plural mangos or mangoes) is a genus of about 35 species of tropical fruiting trees in the flowering plant family Ana cardiaceae, native to India and Southeast Asia, of which the Indian Mango *M indica* is by far the most important commercially. Reference to mangos as the "food of the gods" can be found in the Hindu Vedas. The name of the fruit comes from the Malayalam word *manga,* and popularised by the Portuguese after their Indian exploration, hence the word 'manga' in Portuguese. The mango is one of the ancient fruits of India and its cultivation in India is estimated at more than 4000 to 6000 years old.

Mango is believed to have first originated from the Indo-Myanmar region where it was known to have been cultivated for over 4,000 years. Mango cultivation has now spread throughout the tropics of Africa, Asia and America. Mangoes are an excellent source of vitamin A and C, which normally vary with variety, and maturity of the fruits. Among internationally traded tropical fruits, mango ranks only second to pineapple in quantity and value. World production of mango in 1997 was 22 million tons. India being the world's largest producer, accounting for almost 50% of total world production, China and Mexico with 9% and 6% respectively. Thailand grows more than 100 native mango cultivars.

Mangos are large trees, reaching 35-40 m in height, with a crown radius of 10 m. The leaves are evergreen, alternate, simple, 15-35 cm long and 6-16 cm broad; when young they are orange-pink, rapidly changing to a dark glossy red, then dark green as they mature. The flowers are produced in terminal panicles 10-40 cm long; each flower is small and white with five petals 5- 10 mm long. After the flowers finish, the fruit takes from three to six months to ripen.

The mango fruit is a drupe; when mature, it hangs from the tree on long stems. They are variable in size, from 10-25 cm long and 7-12 cm diameter, and may weigh up to 2.5 kg. The ripe fruit is variably coloured yellow, orange and red, reddest on the side facing the sun and yellow where shaded; green usually indicates that the fruit is not yet ripe, but this depends on the cultivar. When ripe, the unpeeled fruit gives off a distinctive resinous slightly sweet smell. In the center of the fruit is a single flat, oblong stone that can be fibrous or hairless on the surface, depending on cultivar. Inside the shell, which is 1-2 mm thick, is a paper-thin lining covering a single seed, 4-7 cm long, 3-4 cm wide, 1cm thick.

Production

Mangos are produced in 90 countries worldwide on about 8.5 million acres. Production and acreage have increased by 48% in the last decade, as yields have leveled off. Worldwide, yields average about 6800 Ibs/acre, ranging from a few thousand Ibs/acre to > 17,000 Ibs/acre in intensive plantations.

Production of Mango–Top 10 Mango Producing Countries in the World

India	44%	Philippines	3%
China	13%	Indonesia	3%
Thailand	6%	Nigeria	3%
Mexico	5%	Brazil	2%
Pakistan	4%	Egypt	1%

As of 2011 in India, Mango was grown in an area of 2.312 million ha with an annual production of 15.026 million tonnes, which accounted for 50 per cent of the total world production. The state of Uttar Pradesh dominates the mango production in the North, and it is considered the most important fruit in central and eastern part of the state, from Lucknow to Varanasi, which produce some of the finest varieties. In southern India, Andhra Pradesh state is a major producer of mangoes and specialises in making a variety of mango pickles. These pickles are very spicy. It is almost an essential food in Andhra families, where mango pickle production is a household activity in summer. According to statistics collected by the fruit Development Adviser government of India, out of the total fruit average mangoes alone cover about 70% U.P., Tamil Nadu, Karnataka, Bihar and West Bengal lead in Mango growing.

Varieties

The original wild mangos were small fruits with scantbrous flesh, and it is believed that natural hybridization has taken place between *M indica* and *M sylvatica* Roxb. in Southeast Asia. Selection for higher quality has been carried on for 4,000 to 6,000 years and vegetative propagation for 400 years.

Over 500 named varieties (some say 1,000) have evolved and have been described in India. Perhaps some are duplicates by different names, but at least

350 are propagated in commercial nurseries. Naik described 82 varieties grown in South India. L.B. and R.N. Singh presented and illustrated 150 in their monograph on the mangos of Uttar Pradesh (1956). In 1958, 24 were described as among the important commercial types in India as a whole, though in the various climatic zones other cultivars may be prominent locally. Of the 24, the majority are classed as early or mid-season:

Early

'Bombay Yellow' (Bombai) -high quality

'Maida' (Bombay Green)

'Olour' (polyembryonic)-a heavy bearer.

'Pairi' (Paheri, Pirie, Peter, Nadusalai, grape, Raspuri, Goha bunder)

'Safdar Pasand'

'Suvarnarekha' (Sundri)

Mid-Season

'Alampur Baneshan'--high quality but shy bearer

'Alphonso' (Badami, gundu, appas, khader)-high quality

'Bangalora'(Totapuri, collection, kili-mukku, abu Samada in the Sudan)- of highest quality, best keeping, regular bearer, but most susceptible to seed weevil.

'**Banganapally**' (Baneshan, chaptai, Safeda)-of high quality but shy bearer

'Dusehri' (Dashehari, nirali aman, kamyab)-high quality

'Gulab Khas'

'Zardalu'

Mid- to Late-Season

'Rumani' (often bearing an off-season crop)

'Samarbehist' (Chowsa, Chausa, Khajri)-high quality

'Vanraj'

Late

'Fazli' (Fazli maldaj-high quality

Safeda Lucknow

Often Late

'Mulgoa'-high quality but a shy bearer

'Neelum' (sometimes twice a yearj-somewhat dwarf, of indifferent quality, and anthracnose- susceptible.

Contribution to Diet

Mangos are one of the finest fresh fruits in the world, but can be dried, pickled, or cooked as well. Mangos are higher in vitamin C than citrus fruits. Green mangos are the tropical equivalent of green apples–tart, crisp, and somewhat dry, often eaten with salt. They are cooked or used in salads in the tropics. About 25% of mangos are processed into juices, chutneys, sauces, or dried. The large seed can be processed into a flour, and the fat it contains can be extracted and substituted for cocoa butter.

Per capita consumption of mango itself is not known with certainty, but is likely 1-2 pounds per year, based on import data. In consumption data, mango is grouped with other fruits such as avocado, apricot, cherry, kiwi, cranberry and papaya, which collectively have increased about two-fold since 1980 to over 8 lbs per year.

Table Dietary Value, Per 100gram Edible Portion
Food Value per 100gm of ripe Mango flesh

Water (%)	80.0
Calories	63.0
Protein (%)	0.4
Fat (%)	0.4
Carbohydrates (%)	16.0
Crude Fibre (%)	1.0
% of US RDA	
Vitamin A	20.0
Thiamin B_1	3.6
Riboflavin, B_2	2.5
Niacin	2.2
Vitamin C	200.0
Calcium	1.1
Phosphorus	1.5
Iron	4.0
Sodium	--
Potassium	--

* Percent recommended daily allowance set by FDA, assuming a 154 lb male adult, 2700 calories per day.

The fresh kernel of the mango seed (stone) constitutes 13% of the weight of the fruit, 55% to 65% of the weight of the stone. The kernel is a major by product of the mango processing industry. In times of food scarcity in India, the kernels are roasted or boiled and eaten. After soaking to dispel the astringency (tannins), the kernels are dried and ground to flour which is mixed with wheat or rice flour to make bread and it is also used in puddings.

The peel constitutes 20% to 25% of the total weight of the fruit. Researchers in India have shown that the peel can be utilized as a source of pectin. Average yield on a dry weight basis is 13%. Indian analyses of the mango kernel reveal the amino acids–alanine, argentine, aspartic acid, cystine, glutamic acid, glycine, histidine, isoleucine, leucine, lysine, methionine, phenylalanine, proline, serine, threonine, tyrosine, valine at levels lower than in wheat and gluten. Tannin content may be 0.12-0.18% or much higher in certain cultivars.

Preparing Mangoes for Cooking or Eating

Wash mangoes. Place a peeled mango upright on a cutting board, balancing it on the narrow end (Raw mangoes may be peeled with a vegetable peeler; the peel on ripe mangoes may have to be separated after cutting). The seed is large and oval, extending almost through the entire fruit, with the narrow end at the tip of the fruit. Using a sharp knife, slice the mango flesh away on either side of the seed, leaving 3 pieces - 2 mango 'cheeks', and a central seed covered with little flesh. Let the knife 'find' and shape the cutting motion as the flesh is sliced away from the seed. Now, the flesh can be cut into chunks, and any excess flesh around the seed can be trimmed away and used, as well. To prepare a mango for eating, it may be advantageous not to peel at the start, scoop the flesh out of the 'cheeks' with a spoon, for consumption.

Uses

Ripe fruit is used to prepare squash, nectar, jam. Cereal flakes, custard powder and baby food, mango leather (aam papar) and toffee. While, fresh mango remains a delectable treat, the fruit can be used in several other ways. The consistency of the pureed fruit lends itself to being made into many condiments and delectable treats. It can be made into jams and spreads, curds, butters and fruit leathers. For a savory touch, relishes, salsas, sauces, and chutneys are becoming increasingly popular and sought-after items. Mango slices or chunks may also be frozen. Several of these preserved foods can then be used in everyday dishes including shakes, yogurt, ice cream and baked goods.

A simple way to eat a large mango 'as is' involves using a knife. This involves removing part of the skin and then slicing out bite-sized pieces with the knife. More skin can be removed to expose more flesh. One should expect to get juicy hands when eating the last part, when there is no skin to hold with the hand.

Another way to eat a mango is to simply use a sharp knife to peel the skin completely, then make horizontal and vertical cuts on each side through to the flat stone. Then slice the flesh off from each side of the stone. After this slice the remaining flesh left on the side of the stone. This method works best on mangoes that are ripe and which have firm flesh. Another simpler way to enjoy mangos is to buy them frozen or dried as they are becoming more common in this form in local markets.

Mangoes are widely used in chutney, which in the West is often very sweet, but in the Indian subcontinent is usually sharpened with hot limes. In India, ripe mango is often cut into thin layers, desiccated, folded, and then cut and sold as bars that are very tasty and chewy. These bars, known as amavat in Hindi, are similar to dried Guava fruit bars available in Colombia. Many people like to eat unripe mangoes with salt and in regions where food is hotter, with salt and chilly.

Mango is also used to make juices, both in ripe and unripe form. Pieces of fruit can be mashed and used in ice cream; they can be substituted for peaches in a peach (now mango) pie; or put in a blender with milk, a little sugar, and crushed ice for a refreshing beverage. A more traditional Indian drink is mango lassi, which is similar, but uses a mixture of yoghurt and milk as the base, and is sometimes flavoured with salt or cardamom. Dried unripe mango used as a spice in south and southeast Asia is known as amchur (sometimes spelled amchoor). Am is a Hindi word for Mango and amchoor is nothing but powder or extract of Mango.

Ama or Aam is one of the most popular fruit crops in India. In general, the Mangos should always be washed to remove any sap residue, before handling. Some seedling mangos are so fibrous that they cannot be sliced; instead, they are massaged, the stem-end is cut off, and the juice squeezed from the fruit into the mouth. Non-fibrous mangos may be cut in half to the stone, the two halves twisted in opposite directions to free the stone, which is then removed, and the halves served for eating as appetizers or dessert. Or the two "cheeks" may be cut off, following the contour of the stone, for similar use; then the remaining side "fingers" of flesh are cut off for use in fruit cups, etc.

Small mangos can be peeled and mounted on the fork and eaten in the same manner. If the fruit is slightly fibrous especially near the stone, it is best to peel and slice the flesh and serve it as dessert, in fruit salad, on dry cereal, or in gelatin or custards, or on ice cream. The ripe flesh may be spiced and preserved in jars. Surplus ripe mangos are peeled, sliced and canned in sirup, or made into jam, marmalade, jelly or nectar. The extracted pulpy juice of fibrous types is used for making mango halva and mango leather. Mango juice may be spray-dried and powdered and used in infant and invalid foods, or reconstituted and drunk as a beverage. The dried juice, blended with wheat flour has been made into "cereal" flakes. A dehydrated mango custard powder has also been developed in India, especially for use in baby foods.

Ripe mangos may be frozen whole or peeled, sliced and packed in sugar (1 part sugar to 10 parts mango by weight) and quick-frozen in moisture-proof containers. The diced flesh of ripe mangos, bathed in sweetened or unsweetened lime juice, to prevent discoloration, can be quick-frozen, as can sweetened ripe or green mango puree. Immature mangos are often blown down by spring winds. Half-ripe or green mangos are peeled and sliced as filling for pie, used for jelly, or made into sauce, which, with added milk and egg whites, can be converted into mango sherbet. Green mangos are peeled, sliced, parboiled, then combined with sugar, salt, various spices and cooked, sometimes with raisins or other fruits, to make chutney; or they may be salted, sun-dried and kept for use in chutney and pickles. Thin slices, seasoned with turmeric, are dried, and sometimes powdered, and used to impart an acid flavor to chutneys, vegetables and soup. Green or ripe mangos may be used to make relish.

Mango Jam

For some, mangoes are exotic fruits. Cooking the ripe fruit with sugar until they are mush and set, may be an age old technique of fruit recipe for mango jam is simple. Pick ripe mangoes, peel and cut them into small cubes. Add little bit of lime juice to give that extra acidic edge and cook the fruit with sugar until it turns mushy and comes together like firm, yet moist solidity. Store it in a clean glass jar and refrigerate.

Recipe

(for about 1 cup of jam)

- 1 big, ripe mango (peeled and cut into small cubes - 1.5 cup of cut fruit)
- ¼ cup of sugar
- 1 tablespoon of lime juice

Take them all in a clean pot. Cook on medium heat for about 15 minutes, stirring in-between. The sugar will melt and the fruit will break down within 15 to 20 minutes. The soggy, watery mass will come together into solid, moist lump, like *halwa* or *kava* and set. Turn off the heat. Let it cool for few minutes and store in a clean jar.

Mango Bars

Crust

- 2 cups all-purpose flour
- 1/2 cup granulated sugar
- l-cup butter

Filling

- 4 cups chopped mangoes
- 3/4 cup granulated sugar
- 1/3 cup water
- 1 teaspoon fresh lemon juice
- 3 tablespoons cornstarch
- 3 tablespoons water

Topping

- 2 cups quick oats
- 1/4 cup all-purpose flour
- 1/2 cup granulated sugar
- 2/3-cup butter
  - Preheat oven to 350°F.
  - Bake for 7 to 10 minutes or until lightly brown.
  - Filling, combine in a saucepan mangoes, sugar, water and lemon juice. Cook over low-medium heat until mangoes are tender, 10 to 12 minutes. Combine cornstarch and water in a separate bowl. Stir into mango mixture. Cook until mixture is thick. Cool slightly. Pour over prepared crust.
  - Topping, combine oats, flour, and sugar. Cut in butter. Sprinkle over mangoes.
  - Bake for 50 minutes. Cool. Cut into bars.

Mango Shrikhand (Aamrakhand)

Mango pulp (sweetened kesar mango pulp) – 1/2 cup

Plain yogurt - 1 cup

Sour Cream (low-fat is good too) - 1 cup

Finely chopped walnuts and cashews – 1/2 cup (together) Sugar (or Splenda)– 1/2 cup

(Less sugar is okay as the dish gets sweetened from Mango)

Cardamom powder – 1/2 tsp

Saffron soaked in 2 tbsp warm milk - a pinch.

1. Drain the water from the yogurt by tying it in a soft muslin cloth and hang it over the sink for at least 2 hrs.

2. Once all the water is drained from the yogurt, it automatically gets a creamy texture.
3. Mix the yogurt and sour cream thoroughly in a serving dish.
4. Mix in the mango pulp and sugar.
5. Check the sweetness and the flavor; add more sugar or mango pulp if needed.
6. Ensure that the texture of the dish remains creamy and not watery.
7. Mix in the chopped nuts, cardamom powder, and soaked saffron along with the 2 tbsp milk.
8. After mixing thoroughly, chill in the refrigerator for at least 3 hours before serving.

Mango Salsa

"Salsa is a combination of raw, cooked, or partially cooked ingredients that are blended but not cooked together,"

Recipe

- 1 ripe mango - peeled and finely cubed
- 1 fresh corn - grilled and the kernels sliced
- 1 medium red onion or shallot - finely chopped (and washed in water)
- 2 to 4 green chillies - minced Few sprigs of cilantro - finely chopped
- 1 lime - juice squeezed and
- 1/4 teaspoon of salt or to taste
- Toss them together and serve with tortilla (corn flour) chips.

Mango Shake Recipe

Ingredients

Squeezed mangoes or mango pulp – 1/4 of total quantity desired

Milk (preferably not Skimmed) – 1/2 of total quantity desired

Vanilla ice cream or frozen yogurt –1/4 of total quantity desired Cardamom– about 1 piece per cup

Method

Peel the Cardamom and powder the grains; throwaway the peels. Alternatively you can use half a spoon of readily available powdered Cardamom. Add ice cream/ frozen yogurt, milk, and mango pulp to the blender and blend them for about 2 minutes. Add Cardamom to the milk shake and blend it again for a few seconds. Refrigerate the milk shake and serve with or without ice.

A perfect mango shake should have an intense fresh mango flavor and perfume, possess an incredibly deep natural color, have the right consistency (a bit thick at the start, a bit thinner as the ice melts), just the right amount of sweetness and be ice cold on arrival. Next make some sugar water by dissolving 1 cup of sugar in 1/2 cup of water over low heat. Allow this mixture to cool. To make, pull out a really good blender. Pile in the flesh of four mangoes, lots of ice, about 114 or less of your sugar water and some cold water (1/4 cup to start) and blitz away. Test the mixture for thickness and adjust by adding sugared water or water as you please.

Mango Sauce

- 5.5 cups or 3.25 pounds mango puree (use slightly under-ripe to just-ripe mango) (from about 5 pounds, or 5 to 6 whole, large, non-fibrous mangoes, as purchased)
- 6 tablespoons honey
- 4 tablespoons bottled lemon juice
- 1/4 cup sugar
- 2.5 teaspoons ascorbic acid
- 1/8 teaspoon ground cinnamon
- 1/8 teaspoon ground nutmeg

Store in a dark place, away from direct light, to preserve the color of the canned sauce. This sauce is best used within 4 to 6 months; otherwise, discoloration may occur.

Procedure

1. Wash and rinse half-pint canning jars; keep hot until ready to use. Prepare lids according to manufacturer's directions.
2. Wash, peel, and separate mango flesh from seed. Chop mango flesh into chunks and puree in blender or food processor until smooth.
3. Combine all ingredients and heat on medium-high heat, with continuous stirring, until the mixture reaches 200°F. The mixture will sputter as it is being heated, so be sure to wear gloves or oven mitts to avoid burning skin.
4. Fill hot sauce into clean, hot half-pint jars, leaving 1/4-inch head space. Remove air bubbles and adjust head space if needed. Wipe rims of jars with a dampened, clean paper towel; adjust two-piece metal canning lids.

Mango Leather

- 4 cups mango puree (from about 4 large, unripe mangoes)
- 1 cup clover honey

- 1/2 teaspoon ground cinnamon
- 1/4 teaspoon ground nutmeg
- 1/4 teaspoon ground cloves

Yield: About 2 dryer trays (14 inches in diameter); 8 fruit rolls.

Procedure

1. Preheat electric dehydrator to 140°F.
2. Wash and peel mangoes, chop roughly into chunks. Puree in blender until smooth. Pass puree through a food mill or sieve; discard any coarse fiber extracted in food mill. Add honey and spices to the puree and mix thoroughly.
3. Lightly spray two fruit roll tray liners from an electric dehydrator with vegetable oil cooking spray. Spread mango mixture evenly to 1/4-inch thickness on the trays.
4. Position fruit roll liners on dryer trays and place in dehydrator. Dry continuously for about 10 hours. *Maintain dehydrator air temperature steadily at 140°F.* (Monitor the dehydrator air temperature periodically with a thermometer). Remove trays from dehydrator when puree is dry, with no sticky areas (about 10 hours - this will be highly dependent on the relative humidity of the drying room). Test for dryness by touching gently in several places near center of leather; no indentation should be evident.
5. Peel leather from trays while still warm. Leave the second tray on the dehydrator while peeling the first leather, or re-warm leathers slightly in the dehydrator if they cool too much prior to peeling. Cut into quarters; lay on a piece of clean plastic food storage wrap about 1 to 2 inches longer at each end of the leather and roll together into fruit leather rolls. When cool, twist the ends of the plastic wrap tightly to close.
6. Store fruit rolls in freezer-quality zippered plastic bags or airtight plastic container for short-term storage, up to about I month. Leathers should be stored in a cool, dark, dry place. For longer storage up to 1 year, place tightly wrapped rolls in the freezer.

Mango Chutney

- 11 cups of chopped unripe (hard) mango, either (about 9 to 10 large whole mangoes)
- 2.5 cups or 3/4 pound finely chopped yellow onion (about 1 pound onions as purchased)
- 2.5 tablespoons grated fresh ginger

- 0.5 tablespoons finely chopped fresh garlic
- 0.5 cups sugar
- 3 cups white distilled vinegar (5%)
- 2.5 cups golden raisins
- 1.5 teaspoon canning salt
- 4 teaspoons chili powder

Procedure

1. Wash and rinse pint or half-pint canning jars; keep hot until ready to use. Prepare lids according to manufacturer's directions.
2. Wash all produce well. Peel, core and chop mangoes into 3/4-inch cubes. Chop mango cubes in food processor, using 6 one-second pulses per food processor batch. (Do not puree or chop too finely.) By hand, peel and dice onion, finely chop garlic, and grate ginger.
3. Mix sugar and vinegar in an 8- to 10-quart stockpot. Bring to a boil, and boil 5 minutes. Add all other ingredients and bring back to a boil. Reduce heat and simmer 25 minutes, stirring occasionally.
4. Fill hot chutney into clean, hot pint or half-pint jars, leaving 1/2-inch headspace. Remove air bubbles and adjust headspace if needed. Wipe rims of jars with a dampened, clean paper towel; adjust two-piece metal canning lids.

Chapter 12

Papaya (*Carica papaya*)

Papaya (Carica papaya L.), a native of tropical America, has now spread all over the tropical world. Papaya is grown mostly for fresh consumption or for production of proteolytic enzyme papain from the fruit latex (Desai, 1981. In India it is mainly cultivated in the states of Uttar Pradesh, West Bengal, Assam, Gujarat, Bihar, Maharashtra and Madhya Pradesh.

Table : Papaya and Its Related Information

Scientific name	*Distribution in India*	*Production (mmt)*	*Climatic Requirement*	*Soil Requirement*
Carica papaya	All over India to 1500 m above MSNL	1.2	Dry warm climate cannot stand strong wind	All type of soil but best is loamy soil can not grow in water logged clayed soil

Source: Handbook of Agricultural Science 2000.

India is the largest producer of papaya in the world. It is estimated that the present area under papaya is around 20,000 hectares and the productivity of the fruit is 30 *t/ha* in India. It is highly problematic, complicated and interesting fruit crop from botanical, genetical, cytogenetical and horticultural points of view.

Uses

Papaya is used to prepare a number of products. Some of them are listed below :-

Papain

Papain is the dried latex from the fruit, containing a protein - hydrolyzing enzyme, which has a number of specific technological functions. Papain is a hydrlozying enzyme which breaks down proteins into amino acids. It is there fore a protease or proteolytic enzyme. It is an industrial product in preparing various digestive medicine and foods, chill, chill proofing of beer in pharmaceutical industries, textile, garments, cleaning, paper and adhesive manufacture, dental and face cream and swage dispose etc.

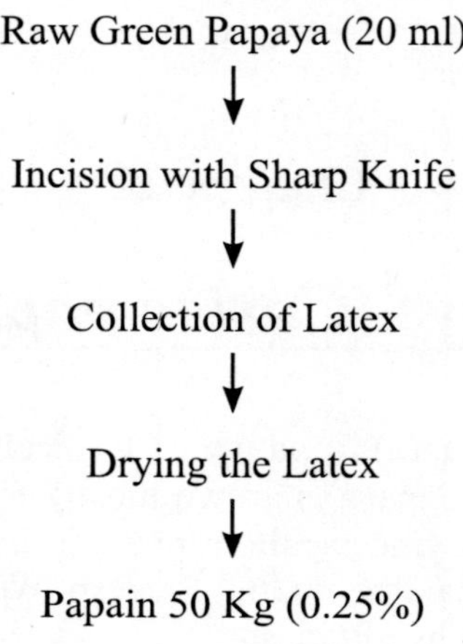

Production of Pectin from Papain

Studies reported that crude papain yield decreased with successive tapping and tapping at 7 day interval resulted in highest latex yield.

Pectin

Pectin is a group of complex materials of very high molecular weight, which are able to form a gel in the presence of correct amount of acidity and sugar.

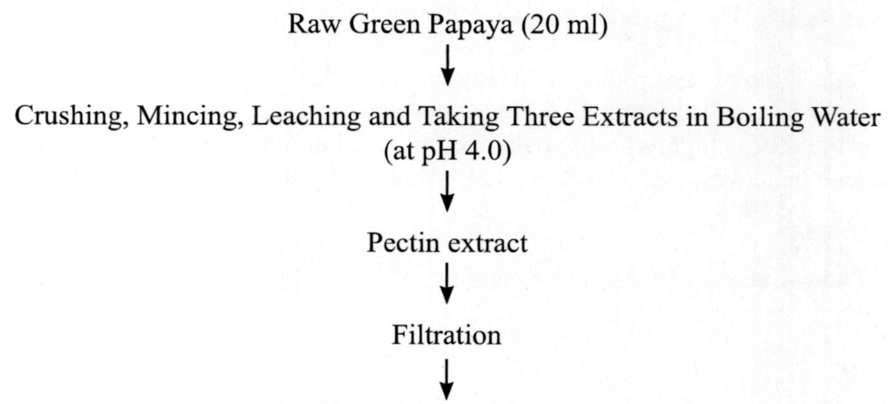

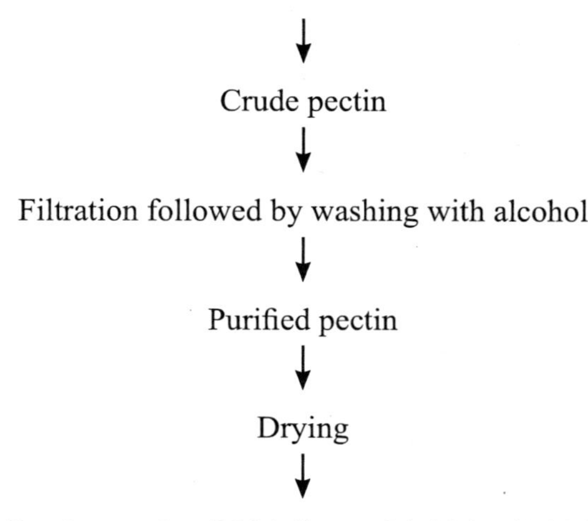

Flowchart Production of Pectin from Papaya

Pickles

Raw papaya can be used for making salted pickles by brine curing and adding spiced vinegar in the traditional way.

Jam

Fully mature and ripe fruit are peeled, deseeded and cut into pieces. These pieces are then cooked with a little water. The cooked pieces are then mashed and sugar added in equal to three fourths of the mashed fruit slices. Citric acid is added at the rate of 5g/kg. The mixture is then cooked with continuous stirring until it attains a thick consistency corresponding to 65-68° brix. The jam is filled hot into clean, sterilized and dry containers. The T.S.S. content of jam is 60°C brix while acidity is 0.72%.

Jelly

Fully developed but somewhat raw fruit are thoroughly cleaned, peeled, deseeded and cut into pieces. They are cooked in water to extract is cooled immediately and allowed to settle for about 2hr. the clarified extract is then decanted and then cooking until a satisfactory sheeting test is obtained. Jelly prepared from papaya was found to have TSS 67% and acidity and acidity 0.85%.

The jelly prepared from green papaya with addition of color was of excellent Quality.

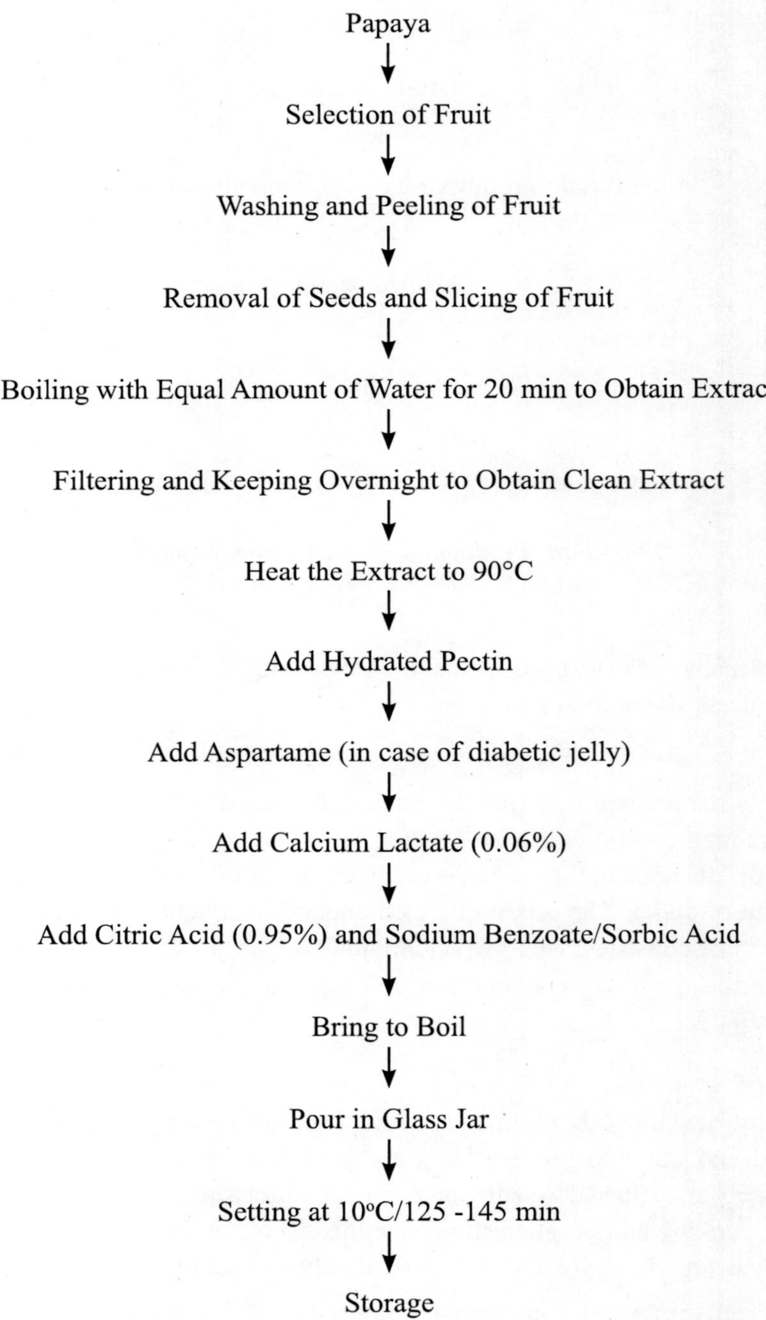

Flowchart Production of Papaya Jelly

Slab

The pulp can also be dehydrated after the addition of sugar (5-7.5%) citric acid (0.5%) and potassium meta-bisulfide (0.3%). It is spread on greased aluminum or stainless steel trays in layers of 1 cm thickness and dried in a cabinet drier at 55-60°C. The dried material, which has leathery consistency, is rolled and cut into pieces of convenient size. The dried product keeps well for about 8 months at 24-30°C.

Concentrate

To make concentrate, papaya puree is treated with 0.05 - 0.2 pectinolytic enzymes at a temperature of 50-60°C to reduce the consistency and kept for 1-2hr. the depectinized puree is concentrated up to threefold in a vacuum concentrator.

Apart from these candy, nectar, puree, tutty fruity, toffee powder, freeze dried chunks, dried rolls and dried slices can be prepared from papaya; Nectar of composition 20% pulp, 13% TSS and 0.3% acidity, Squash of composition 25% pulp, 50% TSS and 1.1 % acidity were found ideal by some workers. Candy prepared with 75% TSS and sauce containing 1 kg pulp, 50 g sugar, 14 salt, 50 g onion, 5 g garlic, 10 g ginger, 5 g red chillies, 10 g hot spices with 4 ml acetic acid (glacial) were reported by researchers the best.

Chapter 13

Ber (Zizyphus Mauritiana Lam.)

Ber is a tropical fruit (also known as Indian jujube), which grows from the tree species, *Ziziphus mauritiana Lamk*. It belongs to the family Rhamnaceae. The tree is 7 - 12 m in height and the trunk, 30 cm in diameter. Branches are slender and downy, bearing paired, brown spines, one straight and the other slightly hooked. It has a wide spreading crown and a short bole. It is a fast growing tree, with an average bearing life of 25 years. The fruits are variable in shape, colour and size. Round, oval or oblong in shape, the fruits may be yellow, green, reddish or purple/ dark brown in colour. Wild fruits are 1.5 - 2.5 cm in length and 1-2 cm in diameter, however improved cultivars can be up to 5 cm in length. The ripe fruit (drupe) is filled with a juicy, hard or soft, sweet-tasting pulp. The flowers are pollinated by honey bees (Apis spp.) and house flies (Musca domestica).

Ber or Indian Jujube (Ziziphus mauritiana) is a tropical fruit tree species, belonging to the family Rhamnaceae. The tree grows very fast even in dry regions reaching heights of 12 meters with a lifespan of 25 years. It is native to Asia (mainly India) though it can also be found in Africa. The ripe fruit, drupe is juicy, soft, and sweet in taste. The fruits ripen at different times even on a single tree and look golden yellow when fully ripe. The size and shape depends on various varieties from the wild to the harvested kind. The fruit is eaten raw or pickled or used in beverages. It is quite nutritious, and rich in vitamin C.

The plant is a vigorous grower and has a rapidly-developing taproot. It may be a bushy shrub 4 to 6 ft (1.2-1.8 m) high, or a tree 10 to 30 or even 40 ft (3-9 or 12 m) tall; erect or wide-spreading, with gracefully drooping branches and downy, zigzag branchlets, thornless or set with short, sharp straight or hooked spines. It may be evergreen, or leafless for several weeks in hot summers. The leaves are alternate, ovate- or oblong-elliptic, 1 to 2.5 inch (2.5-6.25 cm) long, 3/4 to 1.5 inch (2-4 cm) wide; distinguished from those of the Chinese jujube by the dense, silky, whitish or brownish hairs on the underside and the short, downy petioles. On the upper surface, they are very glossy, dark-green, with 3 conspicuous, depressed, longitudinal veins, and there are very fine teeth on the margins.

The 5-petalled flowers are yellow, tiny, in 2's or 3's in the leafaxils. The fruit of wild trees is 1.5 to 1 inch (1.25-2.5 cm) long. With sophisticated cultivation, the fruit reaches 2.5 inch (6.25 cm) in length and 1 3/4 inch (4.5 cm) in width. The form may be oval, obovate, round or oblong; the skin smooth or rough, glossy, thin but tough, turns from light-green to yellow, later becomes partially or wholly burnt-orange or red-brown or all-red.

When slightly underripe, the flesh is white, crisp, juicy, acid or subacid to sweet, somewhat astringent, much like that of a crabapple. Fully ripe fruits are less crisp and somewhat mealy; overripe fruits are wrinkled, the flesh buff-colored, soft, spongy and musky. At first the aroma is applelike and pleasant but it becomes peculiarly musky as the fruit ages. There is a single, hard, oval or oblate, rough central stone which contains 2 elliptic, brown seeds, 1/4 inch (6 mm) long.

Ber (Zizyphus mauritiana) is an important hardy fruit of India, cultivated throughout the country, often called poor man's fruit. It is also considered as *king of fruits of arid regions*. Ber fruits are very nutritious and are rich in vitamins A, B and C. The cultivated Indian ber is very distinct from the Chinese ber, which is less rich in sugar and vitamin C. The Indian ber has a spreading tree, vine-like branches, leaves, which are dark, green on the upper surface. It flowers in autumn, bear's fruits at the end of winter. It sheds the leaves in the hot weather after fruiting and does not like cold climate. The Sikh community also considers the tree sacred. The Golden temple in Amritsar has a '*Ber*' tree called the *'Beri Sahib'* in its central courtyard. The tree is also worshipped.

Varieties

Following varieties are recommended:-

Gujarat:	Umran, Gola, Mehrun Randheri
Haryana:	Gola, Kaithli, Banararsi Karaka, Umran, Seo
Maharashtra:	Shamber, Mehrun, Darakhi, Kharki
Punjab:	Gola, Nazuk, Seo, Sandhura, Narnaul, Kaithli, Banarasi and Umran.
Rajasthan:	Gola, Seb, Jogia, Mundia, Katha (Umran)
Uttar Pradesh:	Narma Varanasi, Delhi Gola, Banarasi Gola, Banarasi Karaka, Muthia, Muriya, Jogia, Aliganj Bihar Nagpuri, Banarasi, Thornless
Andhra Pradesh:	Dudhia, Banarasi

The genus Ziziphus has approximately 40 species, including Ziziphus jujuba (the Chinese jujube). The tree grows widely in most dry tropical and subtropical regions of the world. Its origin is uncertain, though it is thought to be native to tropical Asia, from where it was carried by seafaring traders to Africa and Australia. The tree is well adapted to arid and semi-arid conditions with adequate rain during the vegetative period. It grows in neutral or slightly alkaline, deep, sandy loams, though it can tolerate a range of soil types including those exhibiting moderate salinity. Soils should be well drained, but ber can withstand temporary flooding. The tree has a deep and extensive root system. Fruits and foliage of older trees can be damaged in frost but the trees recover (resprout) even after exposure to temperatures of 60 C.

Ber can provide food security, due to sustained production of the fruit, irrespective of drought, as the tree is drought and saline tolerant and can grow on poor and degraded land. Income from the fruit, fruit products and pruned wood sold for fuel and fencing is therefore consistent throughout the year.

Ziziphus mauritiana, more commonly known as the Chinese apple, is widely cultivated in dry areas throughout the tropics. The Chinese apple has a multitude of uses from culinary to medicinal values and also has the ability to tolerate extremely dry habitats. This makes it an extremely valuable tree for many different cultures that live in such climates. It also has the ability however to form dense stands and become invasive in some areas, including Fiji and Australia. In Australia Ziziphus mauritiana has a capacity to greatly expand its current range in northern and northeastern Australia. The main industry affected is the cattle industry but the species also has environmental impacts in woodland and savanna ecosystems.

Nutrition

The Ber fruit has high sugar content and a high level of vitamins A and C, carotene, phosphorus and calcium. The leaves contain 5% digestible crude proteins, which is an excellent source of ascorbic acid and carotenoids. 100 gm of ripe edible portions contain 80 Calorie. The fruit is especially rich in Vitamin C. A handful of the fruit contain more than a day's requirement of this vitamin. The fruit also contains valuable amounts of phosphorus.

Table : 13.1. Food Value Per 100g of Edible Portion

Fruits, Fresh	
Moisture	81.6-83.0 g
Protein	0.8 g
Fat	0.07 g
Fiber	0.60 g
Carbohydrates	17.0 g
Total Sugars	5.4-10.5 g
Reducing Sugars	1.4-6.2 g

Non-Reducing Sugars	3.2-8.0 g
Ash	0.3-0.59 g
Phosphorus	26.8 mg
Iron	0.76-1.8 mg
Carotene	0.021 mg
Thiamine	0.02-0.024 mg
Riboflavin	0.02-0.038 mg
Niacin	0.7-0.873 mg
Citric Acid	0.2-1.1 mg
Ascorbic Acid	5.8-76.0 mg
Fluoride	0.1-0.2 ppm
Pectin (dry basis)	2.2-3.4%

The fresh fruits also contain some malic and oxalic acid and quercetin.

Fruit Dried	
Calories	73/lb(1,041)
Moisture	68.10 g
Protein	1.44 g
Fat	0.21 g
Carbohydrate	2.47 g
Sugar	21.66 g
Fiber	1.28 g

Uses

The fruit is the most well-known and used product from the ber tree. The fresh fruit has a mild sub-acid flavor and crisp firm flesh, it can also be eaten boiled, as an addition to rice or millet, stewed or baked. Other culinary uses include preparation of pickles, jams, candied fruits, beverages, ber butter and cheese-like pastes. The leaves are used as forage for cattle/sheep/goats and are also palatable for human consumption, used as a vegetable in couscous. The timber, though very hard, can be worked to make fine-grained tools, boats, charcoal and poles for house building. Roots, bark, leaves, wood, seeds and fruits are reputed to have medicinal properties. The tree is also used as a source of tannins, dyes, silk (via silkworm fodder), shellac and nectar. The ber tree can be planted in areas for soil conservation in dune lands, as a windbreak and as living fences for stock control.

The ripe fruit is very nutritious and is consumed raw. It is also used for making candies, pickles and used in desserts. The powdered raw fruit is eaten to cure ulcers, swellings in mouth, diarrhea. The fruit juice is used to treat venereal disorders. The leaves of the tree are used as fodder for the cattle. A paste of the bark helps to cure boils and tumors. The wood is used for construction and for making agricultural implements. The thorny tree makes good fencing The tree is also used as a host for the lac insect, *Kerria lacca* in India.

Ber fruits are very nutritious and are usually eaten fresh In parts of India and North Africa, the leaves of ber are used as nutritious fodder for sheep and goats. The timber is hard, strong, fine-grained and reddish in color and is most often used to make agricultural implements. The branches are used as framework in house construction and the wood makes good charcoal. In addition, this species is used as firewood in many areas. This thorny tree makes good live fencing and is an excellent agroforestry tree to use in hedges. Ber seeds were an ancient Indian cure for the nausea and vomiting during pregnancy. The acid in the fruit made it popular as a remedy for cuts and ulcers. Dried ripe fruit are a laxative. Leaf and root decoction are a folk cure for all kinds of intestinal disorders.

In Ethiopia, the fruits are used to stupefy fish (possibly there is sufficient saponin for this purpose). The leaves are readily eaten by camels, cattle and goats and are considered nutritious. The fruits are applied on cuts and ulcers; are employed in pulmonary ailments and fevers; and, mixed with salt and chili peppers, are given in indigestion and biliousness. The dried ripe fruit is a mild laxative. The seeds are sedative and are taken, sometimes with buttermilk, to halt nausea, vomiting, and abdominal pains in pregnancy. They check diarrhea, and are poulticed on wounds. Mixed with oil, they are rubbed on rheumatic areas. The leaves are applied as poultices and are helpful in liver troubles, asthma and fever and, together with catechu, are administered when an astringent is needed, as on wounds. The bitter, astringent bark decoction is taken to halt diarrhea and dysentery and relieve gingivitis. The bark paste is applied on sores. The root is purgative. A root decoction is given as a febrifuge, taenicide and emmenagogue, and the powdered root is dusted on wounds. Juice of the root bark is said to alleviate gout and rheumatism. Strong doses of the bark or root may be toxic. An infusion of the flowers serves as an eye lotion.

Food Uses

Most Indians prefer to eat ber ripe and out of hand, but some prefer them boiled for a few minutes. Pickles are made from sour ber fruit. In South East Asia, unripe ber fruit are eaten with salt, and refreshing cold drinks are made from the ripe fruit. Powdered sun-dried ripe fruit are an out-of-season specialty in these parts.

In Africa, dried or fermented fruit are made into sour-sweet cakes with an acidic tang. The tender leaves are a leafy vegetable in Indonesia. In South America, a potent brew is made from the ripe fruit. Bees reared on ber flowers produce a light colored honey. Ber is eaten fresh and also processed into delicious candy.

Chapter 14

Apple (*Malus Domestica*)

Delicious and crunchy apple fruit is one of the most popular fruit favored by health conscious, fitness freaks who believe in "health is wealth." This wonderful fruit packed with rich phyto-nutrients that in the true sense are indispensable for optimal health. The antioxidants in apple have many health promoting and disease prevention properties; thus justifying the adage, *"an apple a day keeps the doctor away."*

Apples are obtained from the medium sized tree belonging to the *rosaceae* family. Scientific name: Malus domestica. The apple tree is originated in the mineral rich mountain ranges of Kazakhstan, and is now being cultivated in many parts of the world. Apple fruit features oval or pear shape; and the outer skin has different colors depending upon the cultivar type. Internally, the juicy pulp has off-white to cream in color and has mix of mild sweet and tart taste. The seeds are inedible because of their bitter taste. Several hundred varieties of apples grown in the US and worldwide either for just eating or as a desert fruit to cooking or baking in many recipes.

Health Benefits of Apple

- Delicious and crunchy apple is one of the popular fruit that contain an impressive list of essential nutrients, which are required for normal growth and development and overall nutritional well-being.
- Apples are low in calories; 100 g of fresh fruit slices provide only 50 calories. The fruits are however, contain no saturated fats or cholesterol; but rich in dietary fiber, which helps, prevent absorption of dietary LDL cholesterol in the gut. The dietary fibers also help protect the mucous membrane of the colon from exposure to toxic substances by binding to cancer causing chemicals in the colon.

- Apple fruit contains good quantities of *vitamin-C* and *beta-carotene.* Vitamin C is a powerful natural antioxidant. Consumption of foods rich in vitamin C helps body develop resistance against infectious agents and scavenge harmful, pro-inflammatory free radicals from the body.
- Apples are rich in antioxidant phyto-nutrients *flavonoids*and *polyphenols.* The total measured anti-oxidant strength (ORAC value) of 100 g apple fruit is 5900 TE. The important flavonoids in apples are quercetin, epicatechin, and procyanidin B_2. Apples are also good in tartaric acidthat gives tart flavor to them. These compounds help body protect from deleterious effects of free radicals.
- In addition, apple fruit is a good source of B-complex vitamins such as riboflavin, thiamin, and pyridoxine (vitamin B-6). Together these vitamins help as co-factors for enzymes in metabolism as well as in various synthetic functions inside the body.
- Apple also contains small amount of minerals like potassium, phosphorus, and calcium. Potassium is an important component of cell and body fluids helps controlling heart rate and blood pressure; thus counters the bad influences of sodium.

Selection and Storage

Fresh apples are readily available in the stores all around the season. Choose fresh, bright, firm textured apples with rich flavor. Avoid fruits with pressure marks over their surface as they indicate underlying mottled of pulp. Fresh apples can be kept at room temperature for few days and stored in refrigerator for two to three weeks. Wash them in clean running cold water before use to remove any surface dust and pesticide/fungicide residues.

Preparation and Serving Tips

Wash apples thoroughly in the running water to remove any surface dust, insecticide/fungicide sprays. Remove the top end using paring knife and cut it into two equal halves. Take out centrally placed small seeds. Slice the fruit into desirable cubes or pieces.

Here are some serving tips:

- Eat apple fruit as they are with skin to get maximum health benefits.
- Sliced apple turns brown (enzymatic brownish discoloration) on exposure to air due to conversion in iron form from *ferrous oxide* to ferric oxide. If you have to serve them sliced, rinse slices in water added with few drops of fresh lemon.

- Cloudy apple juice is a good drink with dinner.
- Apple fruit is also used in the preparation of fruit jam, pie, and fruit salad.

Safety Profile

Good yeild demands close attention and suprevision of apple crop. According to the environmental-working group reports, apple fruit is one of the most heavily pesticide-contaminated produce. The most common pesticides found on apple are organo-phosphorous and organo-chloride pesticides like Permethrin and DDT. Therefore, it is recommended to wash the fruit thoroughly before use.

Apple Fruit (*Malus Domestica*), Fresh, Nutritive Value per 100 g, ORAC Value-5900

Energy	50 Kcal	2.5%
Carbohydrates	13.81 g	11%
Protein	0.26 g	0.5%
Total Fat	0.17 g	0.5%
Cholesterol	0 mg	0%
Dietary Fiber	2.40 g	6%
Vitamins		
Folates	3 µg	1%
Niacin	0.091 mg	1%
Pantothenic acid	0.061 mg	1%
Pyridoxine	0.041 mg	3%
Riboflavin	0.026 mg	2%
Thiamin	0.017 mg	1%
Vitamin A	54 IU	2%
Vitamin C	4.6 mg	8%
Vitamin E	0.18 mg	1%
Vitamin K	2.2 µg	2%
Electrolytes		
Sodium	1 mg	0%
Potassium	107 mg	2%
Minerals		
Calcium	6 mg	0.6%
Iron	0.12 mg	1%

Magnesium	5 mg	1%
Phosphorus	11 mg	2%
Zinc	0.04 mg	0%
Phyto-Nutrients		
Carotene-ß	27 μg	--
Crypto-xanthin-ß	11 μg	--
Lutein-zeaxanthin	29 μg	--

Processing

One medium apple = 3/4 cup of apple juice or 1/2 cup of applesauce 1 1/4 apple = one serving of apple juice (8 oz. or 240 ml) = 120 calories

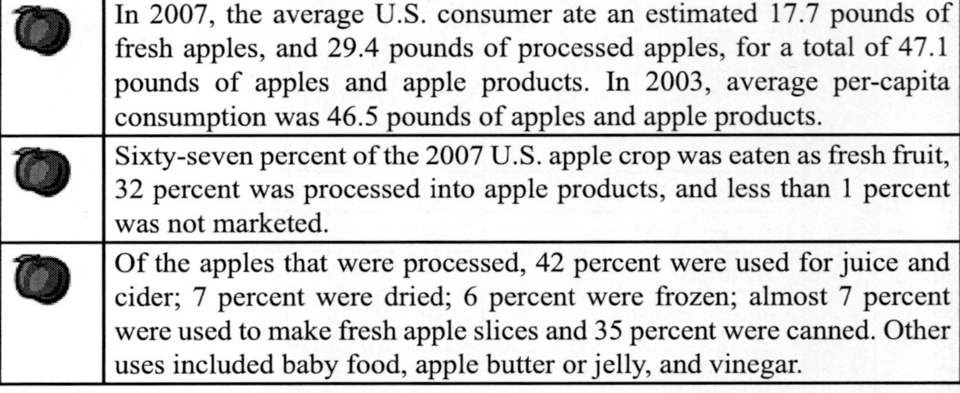

	In 2007, the average U.S. consumer ate an estimated 17.7 pounds of fresh apples, and 29.4 pounds of processed apples, for a total of 47.1 pounds of apples and apple products. In 2003, average per-capita consumption was 46.5 pounds of apples and apple products.
	Sixty-seven percent of the 2007 U.S. apple crop was eaten as fresh fruit, 32 percent was processed into apple products, and less than 1 percent was not marketed.
	Of the apples that were processed, 42 percent were used for juice and cider; 7 percent were dried; 6 percent were frozen; almost 7 percent were used to make fresh apple slices and 35 percent were canned. Other uses included baby food, apple butter or jelly, and vinegar.

Apple Processing

Just like apples that are delivered fresh to your local grocery store, apples that are processed to make apple juice, applesauce and other apple products are picked at their optimum maturity. Only high quality, ripe apples will do! Apples that are an "off" shape or appear to have "skin blemishes" may not be ideal for the produce department - but they are perfectly suitable for processing. As new technologies are developed, the guidelines are updated to allow the industry to produce the best possible products. The first step in any processing procedure is handling of the raw fruit. During this most critical step, there is a visual inspection of all apples by a trained inspector for "integrity and sanitary condition" and random testing for spray residues or mold. Apples not meeting processing standards should be rejected and appropriate personnel informed.

Before raw apples are processed into apple juice, cider or sauce, they are put through a handling process designed to remove external surface dirt and topical chemical residues. These apples are then water-washed before processing. This water wash is sometimes accomplished as the fruit is water flumed from receiving stations to processing lines. Alternately, fruit is transported by dry conveyers through water sprays or scrubbers before processing. Most processing lines employ both techniques.

Water used in the flumes or receiving pits is often recirculated and periodically changed or refreshed. Processors sometimes add chlorine dioxide, hypochlorite or other chlorine compound to control microbial buildup in recirculated water. Apples stay in water flumes or baths for as little as one to two minutes, or as long as 30-45 minutes. Most flumes accomplish apple conveyance to processing lines in less than 10 minutes. Many processors employ high pressure fresh water sprays, sometimes at several points before the fruit enters the processing line. These sprays provide a more vigorous cleaning, and are sometimes used along with mechanical scrubbers, brushes, or bristle rollers to remove surface dirt. Apples are exposed to fresh water sprays for an average of 5 - 10 seconds. Cleaning compounds are not used in water sprays.

The cleaned apples are now ready to be processed into juice. Using various methods, the juice is extracted from the apples and heat-treated (pasteurized) to kill any microorganisms that might be present. This heat treatment also helps improve the overall clarity of the apple juice. Before being placed in the appropriate container (such as bottles), the juice may be further filtered and given an additional heat treatment to assure safety.

Once the apples are cleaned and processed into apple juice or applesauce, they then are subject to analysis using sophisticated, government-approved testing methods that can monitor for even trace amounts of pesticide residues (or other agrichemicals). Apple processors employ strict testing procedures, both in-house and through independent testing laboratories, to ensure the highest quality, purity and safety of their products. Even though apples undergo vigorous cleaning processes, it is possible that some residues may remain after processing. However, the amounts are so small as to be considered insignificant by strict government standards.

Apple processors are committed to providing products that meet the highest possible standards for safety. This is important to know since the health and well-being of both children and adults is paramount to the apple industry and always has been. Apple juice and applesauce are wholesome fruit products that contribute a wide variety of nutrients to the diet.

It also should be known that leading health and scientific organizations such as the American Academy of Pediatrics, American Dietetic Association, the National Institutes of Health, and National Academy of Sciences all agree that a diet rich in fruits and vegetables is the most healthful that children and adults can consume.

Apple Chutney

Good quality apples are washed and cleaned before peeling them. Then they are cut into small pieces. Simultaneously onions and garlic are also cut into small pieces. Then apple pieces are boiled in water and are allowed to soften. Then, other ingredients except sugar and vinegar are mixed with pieces of apples and the mixture is stirred and heated. Subsequently sugar and vinegar are added and heating and stirring is continued till mixture thickens. Finally, chutney is packed in dry glass bottles. A typical process flow chart is as under

Washing and Drying of Apples

↓

Peeling and Cutting of Apples

↓

Chopping of Onions and Garlic

↓

Boiling of Apples

↓

Mixing, Heating & Stirring of Ingredients

↓

Packing

↓

Preserve

Chapter 15

Banana (*Musa acuminata*)

Banana is the common name for herbaceous plants of the genus *Musa* and for the fruit they produce. It is one of the oldest cultivated plants. They are native to tropical South and Southeast Asia, and are likely to have been first domesticated in Papua New Guinea. Today, they are cultivated throughout the tropics. They are grown in at least 107 countries, primarily for their fruit, and to a lesser extent to make fiber, banana wine and as ornamental plants. Its fruits, rich in starch, grow in clusters hanging from the top of the plant. They come in a variety of sizes and colors when ripe, including yellow, purple, and red. Almost all modern edible parthenocarpic bananas come from two wild species – *Musa acuminata* and *Musa balbisiana*. The scientific names of bananas are ***Musa acuminata***, ***Musa balbisiana*** or hybrids ***Musa acuminata* × *balbisiana***, depending on their genomic constitution. The old scientific names *Musa sapientum* and *Musa paradisiaca* are no longer used.

Banana is also used to describe Enset and Fe'i bananas, neither of which belong to the aforementioned species. Enset bananas belong to the genus *Ensete* while the taxonomy of Fe'i-type cultivars is uncertain. In popular culture and commerce, "banana" usually refers to soft, sweet "dessert" bananas. By contrast, *Musa* cultivars with firmer, starchier fruit are called plantains or "cooking bananas". The distinction is purely arbitrary and the terms "plantain" and "banana" are sometimes interchangeable depending on their usage.

The banana fruits develop from the banana heart, in a large hanging cluster, made up of tiers (called *hands*), with up to 20 fruit to a tier. The hanging cluster is known as a bunch, comprising 3–20 tiers, or commercially as a "banana stem", and can weigh from 30–50 kilograms (66–110 lb). In common usage, *bunch* applies to part of a tier containing 3–10 adjacent fruits.

Individual banana fruits (commonly known as a banana or 'finger') average 125 grams (0.28 lb), of which approximately 75% is water and 25% dry matter. There is a protective outer layer (a peel or skin) with numerous long, thin strings (the phloem bundles), which run lengthwise between the skin and the edible inner portion. The inner part of the common yellow dessert variety splits easily lengthwise into three sections that correspond to the inner portions of the three carpels. The fruit has been described as a "leathery berry". In cultivated varieties, the seeds are diminished nearly to non-existence; their remnants are tiny black specks in the interior of the fruit. Bananas are naturally slightly radioactive, more so than most other fruits, because of their high potassium content, and the small amounts of the isotope potassium-40 found in naturally occurring potassium. Proponents of nuclear power sometimes refer to the banana equivalent dose of radiation to support their arguments.

Bananas are a staple starch for many tropical populations. Depending upon cultivar and ripeness, the flesh can vary in taste from starchy to sweet, and texture from firm to mushy. Both the skin and inner part can be eaten raw or cooked. The banana's flavor is due, amongst other chemicals, to isoamyl acetate which is one of the main constituents of banana oil. During the ripening process, bananas produce a plant hormone called ethylene, which indirectly affects the flavor. Among other things, ethylene stimulates the formation of amylase, an enzyme that breaks down starch into sugar, influencing the taste of bananas. The greener, less ripe bananas contain higher levels of starch and, consequently, have a "starchier" taste. On the other hand, yellow bananas taste sweeter due to higher sugar concentrations. Furthermore, ethylene signals the production of pectinase, an enzyme which breaks down the pectin between the cells of the banana, causing the banana to soften as it ripens.

Bananas are eaten deep fried, baked in their skin in a split bamboo, or steamed in glutinous rice wrapped in a banana leaf. Bananas can be made into jam. Banana pancakes are popular amongst backpackers and other travelers in South Asia and Southeast Asia. This has elicited the expression *Banana Pancake Trail* for those places in Asia that cater to this group of travelers. Banana chips are a snack produced from sliced dehydrated or fried banana or plantain, which have a dark brown color and an intense banana taste. Dried bananas are also ground to make banana flour. Extracting juice is difficult, because when a banana is compressed, it simply turns to pulp. Bananas feature prominently in Philippine cuisine, being part of traditional dishes and desserts like *maruya*, *turrón*, and *halo-halo*. Most of these dishes use the Saba or Cardaba banana cultivar. Pisang goreng, bananas fried with batter similar to the Filipino *maruya*, is a popular dessert in Malaysia, Singapore, and Indonesia. A similar dish is known in the United States as banana fritters. Plantains are used in various stews and curries or cooked, baked or mashed in much the same way as potatoes. Seeded bananas (*Musa balbisiana*), one of the forerunners of the common domesticated banana, are sold in markets in Indonesia.

Flower

Banana hearts are used as a vegetable in South Asian and Southeast Asian cuisine, either raw or steamed with dips or cooked in soups, curries and fried foods. The flavor resembles that of artichoke. As with artichokes, both the fleshy part of the bracts and the heart are edible.

Leaves

Banana leaves are large, flexible, and waterproof. They are often used as ecologically friendly disposable food containers or as "plates" in South Asia and several Southeast Asian countries. Especially in the South Indian states of Tamil Nadu, Karnataka, Andhra Pradesh and Kerala in every occasion the food must be served in a banana leaf and as a part of the food a banana is served. Steamed with dishes they impart a subtle sweet flavor. They often serve as a wrapping for grilling food. The leaves contain the juices, protect food from burning and add a subtle flavor. In Tamil Nadu (India) leaves are fully dried and used as packing material for food stuffs and also making cups to hold liquid foods. The dried leaves are called 'Vaazhai-ch- charugu' in Tamil. In Central American countries, banana leaves are often used as wrappers for tamales.

Trunk

The tender core of the banana plant's trunk is also used in South Asian and Southeast Asian cuisine, and notably in the Burmese dish mohinga.

Nutrition and Research

Bananas are an excellent source of vitamin B_6, soluble fiber, and contain moderate amounts of vitamin C, manganese and potassium. Along with other fruits and vegetables, consumption of bananas may be associated with a reduced risk of colorectal cancer and in women, breast cancer and renal cell carcinoma. Banana has a wonderful effect on the skin too. Banana as a rich source of vitamin A, B and E works as a great anti-aging agent. Banana ingestion may affect dopamine production in people deficient in the amino acid tyrosine, a dopamine precursor present in bananas. In India, juice is extracted from the corm and used as a home remedy for jaundice, sometimes with the addition of honey, and for kidney stones. Individuals with a latex allergy may experience a reaction to bananas.

Nutritional Value Per 100 g (3.5 oz)	
Energy	371 kJ (89 kcal)
Carbohydrates	22.84 g
- Sugars	12.23 g
- Dietary fiber	2.6 g
Fat	0.33 g
Protein	1.09 g
Vitamin A equiv.	3 μg (0%)
Thiamine (vit. B_1)	0.031 mg (3%)
Riboflavin (vit. B_2)	0.073 mg (6%)
Niacin (vit. B_3)	0.665 mg (4%)
Pantothenic acid (B_5)	0.334 mg (7%)
Vitamin B_6	0.4 mg (31%)
Folate (vit. B_9)	20 μg (5%)
Choline	9.8 mg (2%)
Vitamin C	8.7 mg (10%)
Calcium	5 mg (1%)
Iron	0.26 mg (2%)
Magnesium	27 mg (8%)

Food Uses

Banana Chips

Principles of Preservation

Under-ripe mature banana or plantain is cut into thin slices and fried to a crisp texture. The slices can be partially dried before frying which removes some of the moisture and makes them more crispy. Frying removes some of the water, gelatinizes the starch, destroys enzymes and micro-organisms and gives a crisp product with a characteristic aroma and taste. The low moisture content inhibits microbial growth and packaging prevents recontamination.

Equipment Needed

- Knives or small fruit slicer
- Plastic buckets or bowls for soaking fruit
- Plastic sieve for draining the soak water
- Drying trays (solar drying)
- Drying cabinet (for assisted drying)

- Large frying pan or wok
- Thermometer
- Polythene bags
- Bag sealer

It is essential to use under-ripe green bananas as these have the correct texture for drying and frying. If plantain is used, select nearly ripe fruit that has stiff and starchy flesh. Ripe and over ripe bananas and plantains should not be used as the texture is too soft to make the chips.

Fried Banana Chips-Process Flow-Chart

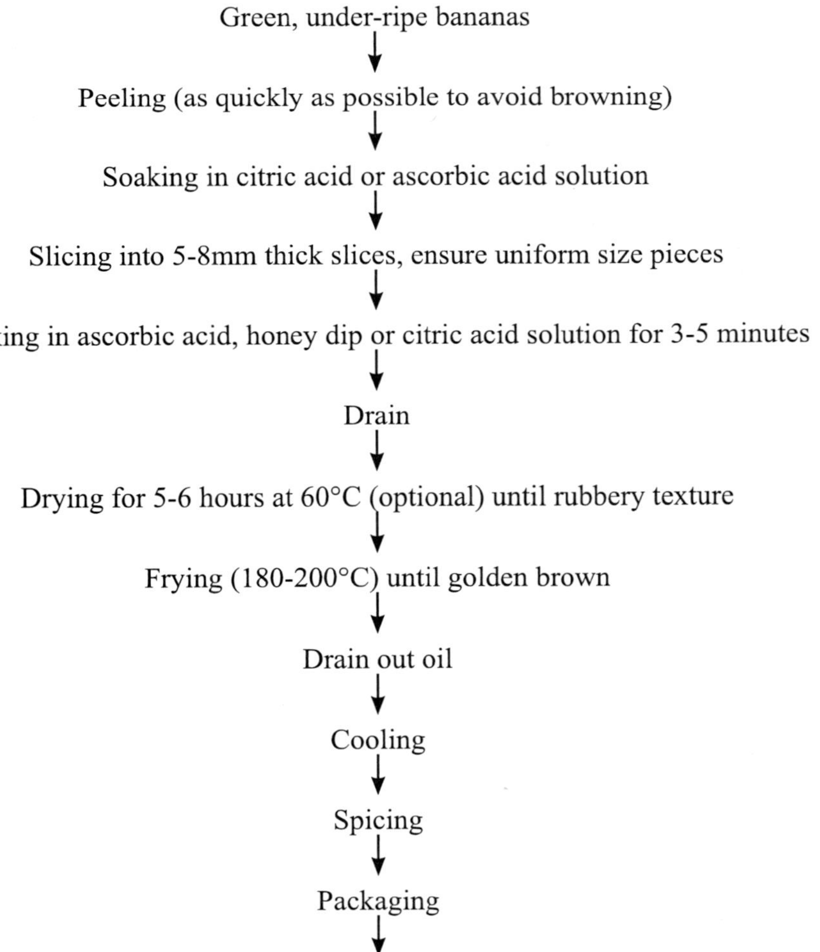

Sun-Dried Banana Chips

Sun-dried banana chips are a popular snack food. They are made by drying the banana pieces on trays under the sun. A solar dryer or cabinet dryer can be used if available–both these interventions will produce higher quality banana chips.

Equipment needed

- Knives or small fruit slicer
- Plastic buckets or bowls for soaking fruit
- Plastic sieve for draining the soak water
- Drying trays (solar drying)
- Drying cabinet (for assisted drying)
- Polythene bags
- Bag sealer

Dried Banana Chips-Process Flow-Chart

Green, under-ripe bananas

↓

Peeling (as quickly as possible to avoid browning)

↓

Soaking in citric acid or ascorbic acid solution

↓

Slicing into 5-8mm thick slices, ensure uniform size pieces

↓

Soaking in honey dip or syrup solution for 3-5 minutes

↓

Drain

↓

Drying for 8 hours at 60°C or 1-2 days in a solar dryer

↓

Packaging

↓

Storage – in a cool place away from direct sunlight up to 2 months

Natural Banana Puree in Aseptic Packing		
1.	***Banana Fruit***	Fresh fruit from the farmer's garden.
2.	***Fruit Receiving***	Raw matured fruit/destemming.
3.	***Fruit Sampling***	Quality approval for firm, matured clean fruit.
4.	***Fruit Ripening***	In atmospheric Controlled ripening Chambers for 4-5 days until suitable uniform ripe fruit is available for processing of pulp.
5.	***Fruit Feeding and Sorting***	Feeding on conveyor belt, sorting out damaged/ Cult fruit and selecting good ripe fruit for processing.
6.	***Fruit Washing***	Removal of dirt, dust and stains.
7.	***Fruit Preparation***	Peeling of the fruit.
8.	***Fruit Pulp Extraction***	Course pulping in feed hopper.
9.	***Homogenization***	To have smooth and homogeneous puree.
10.	***Fine Screening***	Removal of Seeds in turbo refiner.
11.	***Deareation of Puree***	Removal of entrapped air from the puree. By flash cooling under vacuum.
12.	***Quality Lab Analysis***	Recording of brix, acidity, pH, colour, flavour, taste, texture, consistency and viscosity.
13.	***Regen Heating of Puree***	Entering banana puree in aseptic sterilizing heating unit system.
14.	***Final Heating of Puree***	Sterilization of puree to 121°C, by UHT process.
15.	***Holding of Pulp***	Holding the puree in aseptic system at 121°C, for 40 seconds, to ensure the sterility of puree.
16.	***Cooling of Pulp***	Cooling the puree in aseptic system to 20°C to 25°C.
17.	***Aseptic Filling of Pulp***	Filling the puree aseptically in presterillized high barrier aspectic bags in polyethylene liner ms drums.
18.	***Storage***	Store in cool and dry place at a temperature of 5°C to 10°C cold store rooms.

Chapter 16

Pineapple (*Ananas Comosus*)

Pineapple (*Ananas comosus*), a tropical plant with edible multiple fruit consisting of coalesced berries, named for resemblance to the pine cone, is the most economically important plant in the *Bromeliaceae* family. Pineapples may be cultivated from a crown cutting of the fruit, possibly flowering in 20-24 months and fruiting in the following six months.

Pineapple may be consumed fresh, canned, juiced, and are found in a wide array of food stuffs–dessert, fruit salad, jam, yogurt, ice cream, candy, and as a complement to meat dishes. In addition to consumption, in the Philippines the pineapple's leaves are used as the source of a textile fiber called pina, and is employed as a component of wall paper and furnishings, amongst other uses.

The pineapple is a herbaceous perennial which grows to 1.0 to 1.5 meters (3.3 to 4.9 ft) tall, although sometimes it can be taller. In appearance, the plant itself has a short, stocky stem with tough, waxy leaves. When creating its fruit, it usually produces up to 200 flowers, although some large-fruited cultivars can exceed this. Once it flowers, the individual fruits of the flowers join together to create what is commonly referred to as a pineapple. After the first fruit is produced, side shoots (called 'suckers' by commercial growers) are produced in the leaf axils of the main stem. These may be removed for propagation, or left to produce additional fruits on the original plant. Commercially, suckers that appear around the base are cultivated. It has 30 or more long, narrow, fleshy, trough-shaped leaves with sharp spines along the margins that are 30 to 100 centimeters (1.0 to 3.3ft) long, surrounding a thick stem. In the first year of growth, the axis lengthens and thickens, bearing numerous leaves in close spirals. After 12 to 20 months, the stem grows into a spike-like inflorescence up to 15 cm (6 in) long with over 100 spirally arranged, trimerous flowers, each subtended by a bract. Flower colors vary, depending on variety, from lavender, through light purple to red.

The ovaries develop into berries which coalesce into a large, compact, multiple accessory fruit. The fruit of a pineapple is arranged in two interlocking helices, eight in one direction, thirteen in the other, each being a Fibonacci number.

Pineapple carries out CAM photosynthesis, fixing carbon dioxide at night and storing it as the acid malate and then releasing it during the day, aiding photosynthesis. Raw pineapple is an excellent source of manganese (45% DV in a 100 g serving) and vitamin C (80% DV per 100 g).

Mainly from its stem, pineapple contains a proteolytic enzyme, bromelain, which breaks down protein. If having sufficient bromelain content, raw pineapple juice can thus be used as a marinade and tenderizer for meat. Pineapple enzymes can interfere with the preparation of some foods, such as jelly or other gelatin-based desserts, but would be destroyed during cooking and the canning process. The quantity of bromelain in the fruit is probably not significant, being mostly in the inedible stalk. Furthermore, an ingested enzyme like bromelain is unlikely to survive intact the proteolytic processes of digestion.

Nutritional Value per 100 g (3.5 oz)	
Energy	202 kJ (48 kcal)
Carbohydrates	12.63 g
- Sugars	9.26 g
- Dietary fiber	1.4 g
Fat	0.12 g
Protein	0.54 g
Thiamine (vit. B_1)	0.079 mg (7%)
Riboflavin (vit. B_2)	0.031 mg (3%)
Niacin (vit. B_3)	0.489 mg (3%)
Pantothenic acid (B_5)	0.205 mg (4%)
Vitamin B_6	0.110 mg (8%)
Folate (vit. B_9)	15 μg (4%)
Vitamin C	36.2 mg (44%)
Calcium	13 mg (1%)
Iron	0.28 mg (2%)
Magnesium	12 mg (3%)
Manganese	0.9 mg (43%)
Phosphorus	8 mg (1%)
Potassium	115 mg (2%)
Zinc	0.10 mg (1%)

Percentages are relative to US recommendations for adults.
Source: USDA Nutrient Database

Processing

Processed pineapple is a popular product which is exported by countries which produce pineapple. Brazil is considered the main pineapple producing country in the world since 2005. During processing, nutritional quality of pineapple can be reduced but there are recent reasearches carried out which uses new technologies which tries to retain the nutritional quality of the pineapple fruit. This is to meet the consumer demand for healthy, nutritious and "natural" products.

Some of the processing methods using the new technologies are as follows:

- ***Vacuum Frying*** - A dehydration process that produces healthy fruit snacks (pineapple chips) which partially preserve the fruits original colour and nutritional compounds and have a high hydrophilic antioxidant capacity (Perez-Tinoco et al., 2008).
- ***Radiation Processing*** - A dose of 2KGy did not affect significantly the nutritional value as well as the sensory quality of minimally processed pineapple (Hajare et al., 2006).
- ***Thermal Processing*** - helps in the improvement of colour, as a quality attribute of processed pineapple puree. This is made possible by the increase in knowledge of kinetic of colour change (Chutintrasri and Noomhorm, 2007).
- ***Ultrasound*** - This is a pre-treatment for drying of pineapple. Drying of pineapple reduces post-harvest loss of fruits and also a process to produce dried fruits, which can be directly consumed or become part of foodstuffs like cakes, pastries and many others. This method has affected the sugar gain of pineapples during the pre-treatment and also has affected the water effective diffusivity of pineapples during air-drying process (Fernandes et al., 2008).
- ***Osmotic Evaporation*** - This is a process whereby pineapple juice is concentrated at moderate temperatures and pressures with good nutritional and sensory qualities. This process has minor changes in the concentrated juices which makes it more preferable (Hongvaleerat et al., 2008).
- ***High pressure technology*** - This method is used in food processing where food borne micro-organisms and enzymes are inactivated at low temperature, without the need for chemical preservation. This is done in fruit juice processing to preserve most of the nutritional qualities similar to a fresh product (Deliza et al., 2005).

Utilisation of Pineapple Waste

Pineapple waste is a by-product of the pineapple processing industry and it consists of residual pulp, peels and skin. These wastes can cause environmental pollution problems if not utilised. Recently there are investigations/studies carried out on how to utilise these wastes.

Pineapple peel is rich in cellulose, hemicellulose and other carbohydrates. Ensilaging of pineapple peels produces methane which can be used as a biogas. Anaerobic digestion takes place and the digested slurry may find further application as animal, poultry and fish feeds.

Correia et al (2004) investigated the ability of *Rhizopus oligosporous* to produce enhanced levels of free phenolics from pineapple residue in combination with soy flour as potential nitrogen source. From this investigation, they established a relationship between antioxidant activity, ß-glucosidase and total phenolic content in these pineapple/soy flour extracts. They will further investigate these extracts but from this, the value of pineapple wastes can be enhanced.

Pineapple Juice

It is the juice extracted from pineapple that can be marketed by bottling in to sterilized bottles or can. By adding preservative at a specified level (KMS <70ppm or Benzoic acid < 120 ppm), the shelf life of the juice can be extended to 6-7 months. Flow chart for preparation of pineapple juice is shown below:

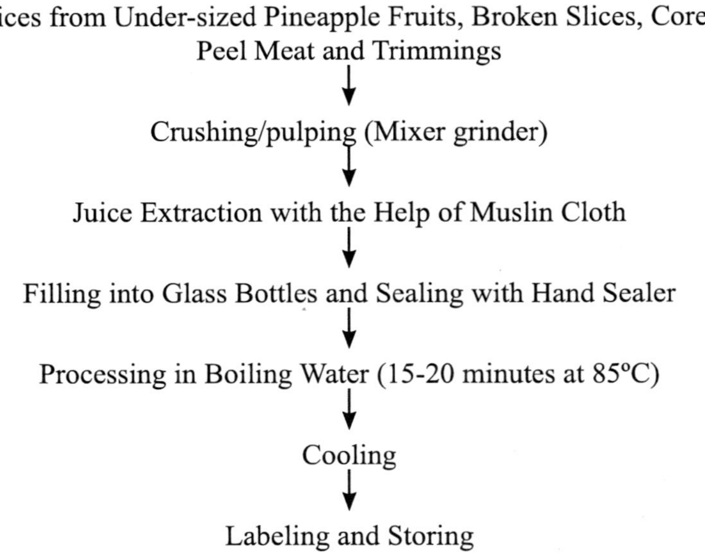

Pineapple Squash

Pineapple squash should be prepared from fully matured and sound pineapple fruits free from insect infestation, diseases etc. For preparation of pineapple squash required quantity of juice, sugar, citric acid, preservative (Potassium metabisulphite or sodium benzoate), water, essence and colour are calculated as per FPO specifications. It should be diluted 2-3 times with water at the time of consumption. Flow chart for preparation of pineapple squash is shown below:

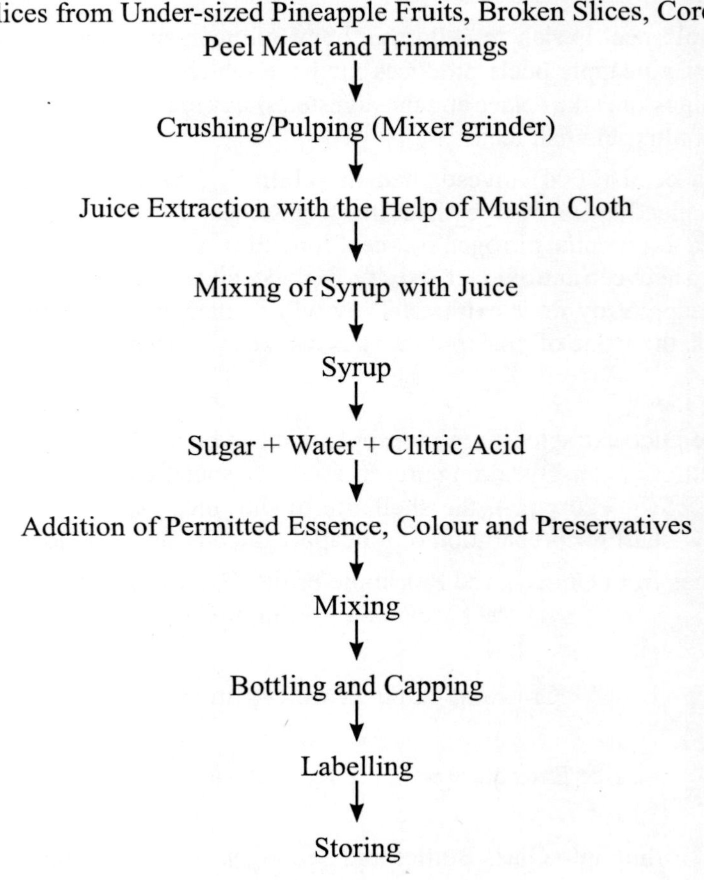

Recipe for Pineapple Squash

1. Pineapple Juice 1.0 kg
2. Sugar 900g-1.2g
3. Water 1.3 liter
4. Citric acid 40-45 g
5. Pineapple Colour 0.5 g
6. Pineapple essence 5-10 ml
7. Potassium meta bisulphite (KMS) Not more than 1.0 g/kg

Ready-to-Serve (RTS) Beverage

The pineapple RTS beverage is prepared from the extracted pineapple juice, adjusting its soluble solids and acidity as per FPO specifications for RTS beverage by mixing the juice with required quantity of sugar syrup prepared from sugar, citric acid and water. Colour and essence as per the requirements are also added and mixed thoroughly. The beverage is filled into bottles leaving a head space of 2.5 to 3.0 cm, crown corked and processed in water for 15-20 mins at 85ºC and air cooled. Flow chart for preparation of Ready-to-Serve Beverage is shown below:

Slices from under-sized pineapple fruits, broken slices, cores, peel meat and trimmings

↓

Crushing/pulping (Mixer grinder)

↓

Juice extraction with the help of muslin cloth

↓

Mixing of syrup with juice ← Syrup ← Sugar + Water + Citric Acid

↓

Addition of permitted essence, & colour

↓

Mixing

↓

Bottling and Sealing

↓

Processing in boiling water (15-20 minutes at 85oC)

↓

Storing

Recipe for Pineapple (RTS) Beverage

1. Pineapple Juice 1.0 kg
2. Sugar 600-700g

3. Water 4.0 liter
4. Citric acid 10-15 g
5. Pineapple Colour 1.5 g
6. Pineapple essence 15-20 ml

Pineapple Jam

Pineapple jam is solid gel made from the fruit pulp or juice, sugar, and pectin. Different steps which are essential during this product preparation are shown below:

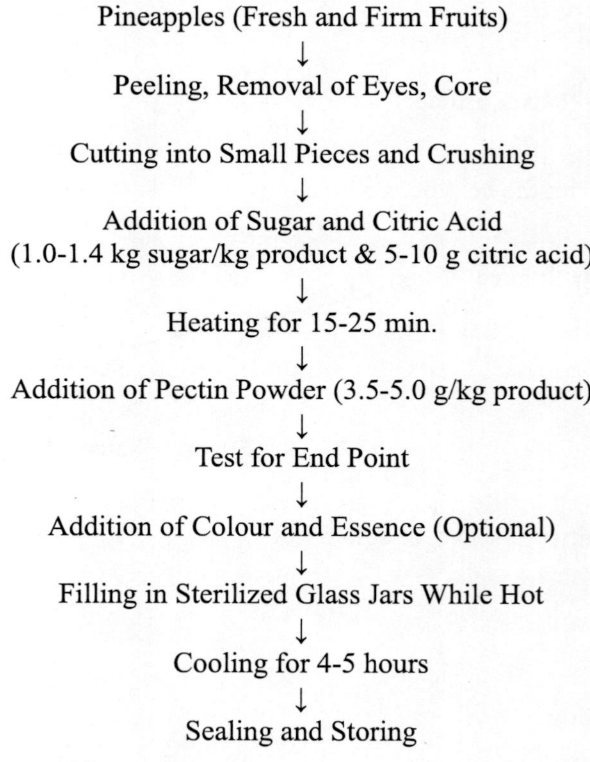

Recipe for Pineapple Jam

1. Pineapple Pulp 1.0 kg
2. Sugar 1.0-1.4g
3. Citric acid 10-15 g
4. Pectin powder 3.5-5.0 g/kg
5. Pineapple Colour 1.5 g
6. Pineapple essence 5-10 ml

Chapter 17

Orange (*Citrus Sinensis*)

An orange—specifically, the sweet orange—is the citrus *Citrus sinensis* (*Citrus sinensis* (L.) Osbeck) and its fruit. It is the most commonly grown tree fruit in the world. The orange is a hybrid of ancient cultivated origin, possibly between pomelo (*Citrus maxima*) and mandarin (*Citrus reticulata*). It is an evergreen flowering tree generally growing to 9–10 m in height (although very old specimens have reached 15 m). The leaves are arranged alternately, are ovate in shape with crenulate margins and are 4–10 cm long. The orange fruit is a hesperidium, a type of berry. Oranges are round citrus fruits with finely-textured skins that are, of course, orange in color just like their pulpy flesh; the skin can vary in thickness from very thin to very thick. Oranges usually range from approximately two to three inches in diameter.

Propagation

Propagation of orange trees is deceptively difficult, because hardy edible oranges are not generally grown from seed. Cultivars that produce good quality fruit are highly susceptible to root diseases. Grafted trees also begin bearing fruit many years earlier than trees reproduced by seed. Other benefits of grafting include more accurate reproduction of good fruit traits than plants derived from seed, and the opportunity to alter tree size, productivity, and other traits through rootstock influence, while maintaining identical fruit characteristics. Almost all orange trees are propagated in two stages. First, rootstock is grown from seed. When the seedling is well-established, the leafy top is cut off, and budwood from an existing tree is grafted onto the rootstock. It is the budwood that determines the variety of orange that is grown.

Climate

Oranges can be grown outdoors in warmer climates, and indoors in cooler climates. Like most citrus plants, oranges will not do well unless kept between

15.5°C–29°C (60°F–85°F). Orange trees grown from the seeds of a store-bought fruit may not produce fruit, and any fruit that is produced may be different than the parent fruit, due to modern techniques of hybridization. To grow the seed of a store-bought orange, one must not let the seed dry out (an approach used for many citrus plants). One method is to put the seeds between the halves of a damp paper towel until they germinate, and then plant them. Many just plant them straight into the soil, making sure to water them regularly. Oranges require a huge amount of water and the citrus industry in the Middle East is a contributing factor to the desiccation of the region.

Oranges are sensitive to frost, and a common treatment to prevent frost damage when sub-freezing temperatures are expected is to spray the trees with water, since as long as unfrozen water is turning to ice on the trees' branches, the ice that has formed stays just at the freezing point, giving protection even if air temperatures have dropped far lower.

Food Uses

Oranges can be eaten as a snack--just peel and enjoy. Before cutting the orange in half horizontally through the center, wash the skin so that any dirt or bacteria residing on the surface will not be transferred to the fruit. Proceed to cut the sections into halves or thirds, depending upon your personal preference.

Thin-skinned oranges can be easily peeled with your fingers. For easy peeling of the thicker skinned varieties, first cut a small section of the peel from the top of the orange. You can then either make four longitudinal cuts from top to bottom and peel away these sections of skin, or starting at the top, peel the orange in a spiral fashion.

Oranges are oftentimes called for in recipes in the form of orange juice. As oranges, like most citrus fruits, will produce more juice when warmer, always juice them when they are at room temperature. Rolling the orange under the palm of your hand on a flat surface will also help to extract more juice. The juice can be extracted in a variety of ways. You could either use a juicer or do it the old fashioned way, squeezing by hand.

If your recipe calls for orange zest, make sure that you use an orange that is organically grown since most conventionally grown fruits will have pesticide residues on their skin and may be artificially colored. After washing and drying the orange, use a zester, paring knife or vegetable peeler to remove the zest, which is the orange part of the peel. Make sure not to remove too much of the peel as the white pith underneath is bitter and should not be used. The zest can then be more finely chopped or diced if necessary.

Nutritional Value of Oranges per 100 g (3.5 oz)	
Energy	192 kJ (46 kcal)
Carbohydrates	11.54 g
- Sugars	9.14 g
- Dietary fibre	2.4 g
Fat	0.21 g
Protein	0.70 g
Thiamine (vit. B_1)	0.100 mg (9%)
Riboflavin (vit. B_2)	0.040 mg (3%)
Niacin (vit. B_3)	0.400 mg (3%)
Pantothenic acid (B_5)	0.250 mg (5%)
Vitamin B_6	0.051 mg (4%)
Folate (vit. B_9)	17 μg (4%)
Vitamin C	45 mg (54%)
Calcium	43 mg (4%)
Iron	0.09 mg (1%)
Magnesium	10 mg (3%)
Phosphorus	12 mg (2%)
Potassium	169 mg (4%)
Zinc	0.08 mg (1%)

Orange Peel

Although not as juicy or delicious as the inside of an orange, the peel is edible, and has been consumed particularly in environments where there is scarcity of resources and where maximum nutritional value must be derived and minimal waste generated (for example, on a submarine). The peel of an orange has increased vitamin C and fiber. However, high concentrations of pesticides have been found in orange peels. Some organizations recommend that one should only consume the peels of organically grown and processed oranges, where chemical pesticides or herbicides would not have been used on the peel. Orange peel contains citral, an aldehyde that antagonizes the action of vitamin A. Therefore, anyone eating orange peels should make certain that their dietary intake of Vitamin A is sufficient

Juice and Other Products

Octyl acetate is responsible for the fragrance of oranges. Oranges are widely grown in warm climates worldwide, and the flavours of oranges vary from sweet to sour. The fruit is commonly peeled and eaten fresh, or squeezed for its juice. It has a thick bitter rind that is usually discarded, but can be processed into animal feed by removal of water, using pressure and heat. It is also used in certain recipes as flavouring or a garnish. The outer-most layer of the rind can be grated or thinly veneered with a tool called a zester, to produce orange zest. Zest is popular in cooking because it contains the oil glands and has a strong flavour similar to the fleshy inner part of the orange. The white part of the rind, called the *pericarp* or *albedo* and including the pith, is a source of pectin and has nearly the same amount of vitamin C as the flesh.

Products Made from Oranges

- Brazil is the largest producer of orange juice in the world, followed by the USA. It is made by squeezing the fruit on a special instrument called a "*juicer*" or a "*squeezer*." The juice is collected in a small tray underneath. This is mainly done in the home, and in industry is done on a much larger scale.
- Frozen orange juice concentrate is made from freshly squeezed and filtered orange juice.
- Sweet orange oil is a by-product of the juice industry produced by pressing the peel. It is used as a flavouring of food and drink and for its fragrance in perfume and aromatherapy. Sweet orange oil consists of about 90% d-limonene, a solvent used in various household chemicals, such as to condition wooden furniture, and along with other citrus oils in grease removal and as a hand-cleansing agent. It is an efficient cleaning agent which is promoted as being environmentally friendly and preferable to petroleum distillates.

However, d-Limonene is classified from slightly toxic to humans to very toxic to marine life in different countries. Its smell is considered more pleasant by some than those of other cleaning agents. Although once thought to cause renal cancer in rats, limonene now is known as a chemopreventive agent with potential value as a dietary anti-cancer tool in humans. There is no evidence for carcinogenicity or genotoxicity in humans. The carcinogenic potency project estimates that it causes human cancer on a level roughly equivalent to that caused by exposure to caffeic acid via dietary coffee intake.

- The orange blossom is highly fragrant and traditionally associated with good fortune. It has long been popular in bridal bouquets and head wreaths for weddings.
- Orange blossom essence is an important component in the making of perfume.
- The petals of orange blossom can also be made into a delicately citrus-scented version of rosewater; orange blossom water (aka orange flower

water) is a common part of both French and Middle Eastern cuisines, most often as an ingredient in desserts and baked goods.

- In some countries orange flower water is used to make orange blossom scones and marshmallows.
- The orange blossom gives its touristic nickname to the *Costa del Azahar* ("Orange-blossom coast"), the Castellon seaboard.
- In Spain, fallen blossoms are dried and then used to make tea.
- Orange blossom honey, or actually citrus honey, is produced by putting beehives in the citrus groves during bloom, which also pollinates seeded citrus varieties. Orange blossom honey is highly prized, and tastes much like orange.
- Marmalade, a conserve usually made with Seville oranges. All parts of the orange are used to make marmalade: the pith and pips are separated, and typically placed in a muslin bag where they are boiled in the juice (and sliced peel) to extract their pectin, aiding the setting process.
- Orange peel is used by gardeners as a slug repellent.
- Orange leaves can be boiled to make tea.
- Orange wood sticks (also spelt orangewood) are used as cuticle pushers in manicures and pedicures, and as spudgers for manipulating slender electronic wires
- Orange wood is a flavouring wood in meat grilling much as mesquite, oak, pecan and hickory are used.

Chapter 18

Tomato (*Lycopersicon esculentum*)

Tomato (*Lycopersicon esculentum*) is a member of the solanacea - the potato family. It is a watery fruit containing 5-7% dry matter. It contains relatively low concentrations of Vitamin C, beta-carotene and minerals Tomato rank fourth among the leading vegetable of the world. World wide about 6-9 million metric tonne of tomatoes are produced annually and the leading tomato producing countries are United Sates, U.S.S.R, Turkey, China, Italy, Egypt, Spain, Romania, Brazil and Greece.

Tomato *(Lycopersicon esculentum)* was first cultivated in Mexico and later travelled to Europe. The Italians called it pomi d'oro (golden apple) and the French pomme d'amour (love apple) giving it special place in their cuisines. It then came to Asia and captured the minds and cuisines of all. No gravy is complete without the addition of this vegetable-fruit. Tomatoes belong to family of potato and eggplant, Solanacaea. Earlier it was considered poisonous due to its association with the nightshade family. Technically it is a fruit (ovary with seeds), but considered a vegetable due to its uses.

Picking and Ripening

Tomatoes are often picked unripe, and ripened in storage with ethylene. Ethylene is the plant hormone produced by many fruits and acts as the cue to begin the ripening process. These tend to keep longer, but have poorer flavor and a mealier, starchier texture than tomatoes ripened on the plant. They may be recognized by their color, which is more pink or orange than the ripe tomato's deep red.

Recently, stores have begun selling "tomatoes on the vine" which are ripened still connected to a piece of vine. These tend to have more flavor (at a price premium) than artificially ripened tomatoes, but still may not be the equal of local garden produce. Also relatively recently, slow-ripening cultivars of tomato have been developed by crossing a non-ripening variety with ordinary tomato cultivars. Cultivars were selected whose fruits have a long shelf life and at least reasonable flavor.

Modern Uses of Tomatoes

Tomatoes are now eaten freely throughout the world. Today, their consumption is believed to benefit the heart. Lycopene, one of nature's most powerful antioxidants, is present in tomatoes and has been found to be beneficial in preventing prostate cancer, among other things. Botanically a fruit, the tomato is generally thought of and used as a vegetable: it's more likely to be part of a sauce or a salad than eaten whole as a snack, let alone as part of a dessert (though, depending on the variety, they can be quite sweet, especially roasted).

Tomatoes are used extensively in Mediterranean and Middle Eastern cuisines, especially Italian ones. The tomato has an acidic property that is used to bring out other flavors. This same acidity makes tomatoes especially easy to preserve in home canning as tomato sauce or paste. Tomato juice is often canned and sold as a beverage. Unripe green tomatoes can also be used to make salsa, be breaded and fried, or pickled.

Tomatoes are also a popular "non-lethal" throwing weapon in mass protests, and there is a common tradition of throwing rotten tomatoes at bad actors or singers on a stage although this tradition is more symbolic as of today. Because to be most known by tomatoes growth and production in Mexico, the Mexican state of Sinaloa takes like symbol the tomato.

Culinary uses of tomatoes include:

- Tomato paste
- Tomato puree
- Tomato pie
- Ketchup
- Pizza
- Spaghetti (Italian cuisine)

Tomato Cocktail

Tomato cocktail is gaining popularity in many of the high-class hotels and restaurants. Although the recipes vary, the main constituent is tomato juice to which common salt, vinegar, lemon etc are added.

Recipe

- Tomato Juice-1L
- Salt-8g

- Sugar (optional)-20 g
- Cloves (headless) powder
- Coriander, large cardamom, cumin, cinnamon, black pepper (all in powder form)-0.3 g each Red chilly powder-0.05 g
- Lime juice-10 ml
- Vinegar (10% acetic acid) - 50 ml.

The method of preparation is as follows

Tomato Juice (Strained)

↓

Putting spice bag in juice and cooking gently for 20 minutes in covered vessel

↓

Adding strained limejuice, vinegar and salt (if desired, a little sugar may also be added)

↓

Heating at 82-88°C

↓

Filling hot in bottles

↓

Crown corking

↓

Storing at ambient temperature (in cool and dry place)

Flow Sheet for Tomato Cocktail

Tomato Chutney

Chutneys are usually prepared in Indian homes. Chutney of good quality should be palatable and appetizing. According to FPO, any fruit chutney should have minimum fruit percentage of 40 percent and TSS 50 percent.

Recipe

- Tomato -1 kg,
- onion (chopped) - 200 g
- Ginger (choppedj-1 0 g
- Garlic (chopped) - Sg
- Red chilly powder - 8g

- Sugar-600 g, salt 30 g

Cinnamon, black pepper, cardamom (large), cumin (all in powder form)-10 g each Glacial acetic acid-5ml, Sodium benzoate-0.5 per kg of the final product Preservative should be added after dissolving in small quantity of water. The method of preparation is as follows :

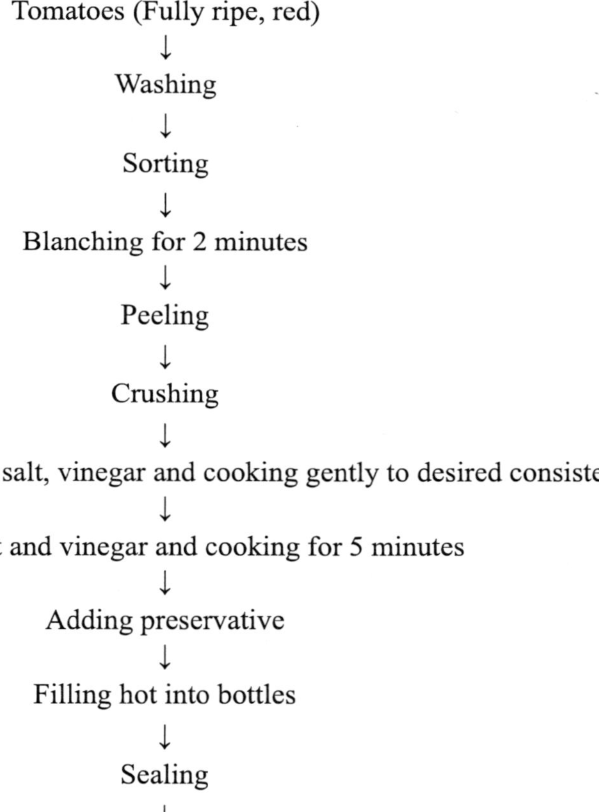

Tomatoes (Fully ripe, red)
↓
Washing
↓
Sorting
↓
Blanching for 2 minutes
↓
Peeling
↓
Crushing
↓
Adding ingredients except salt, vinegar and cooking gently to desired consistency
↓
Adding salt and vinegar and cooking for 5 minutes
↓
Adding preservative
↓
Filling hot into bottles
↓
Sealing
↓
Storing at ambient temperature (in cool and dry place)

Flow Sheet for Tomato Chutney

Whole Tomato Concentrate (Crush)

It is distinct type of product, which contains the whole of tomato inclusive of skin and seeds, and it thus differs from tomato puree and paste, which are concentrated tomato juice products without skin and seeds. The use of crush will give the full complement of whole tomatoes. This is an intermediate product, which can be used for culinary purpose, for soup and chutney preparation.

Recipe

- ➢ Tomato pulp (inclusive of seed and peels) - 1 kg
- ➢ Glacial acetic acid - 1.5 ml
- ➢ Sodium benzoate -70mg
- ➢ Potassium metabisulphite-140 mg.

The method of preparation is as follows

Selecting fully ripe, red and undamaged tomatoes

↓

Washing, removing green portion and cutting into pieces

↓

Boiling and crushing the pieces for uniform pulp

↓

Continue boiling on mild heat till the weight of whole mass is reduced to one third, i.e. into a thick paste

↓

Towards the end, adding a teaspoon (5 ml) of glacial acetic acid for every kg of the paste and boiling for 5-8 minutes

↓

For every kg of finished product adding 0.4 g of potassium metabisulphite and 0.2 g odium benzoate after dissolving in a small quantity of water and mixing thoroughly

Filling the hot crush in clean dry glass jars upto the brim. Sealing the jars tightly and storing in a cool and dry place.

Flow Sheet for Preparation of Whole Tomato Crush

Bibliography

Allen Susser (1997). The Great Citrus Book: A Guide with Recipes. Ten Speed Press. ISBN 978-0-89815-855-7.

Anderson, J. W., Hanna, T. J., Peng, X. and Kryscio, R. J. (2000). Whole Grain Foods and Heart Disease Risk. *J Am Coll Nutr June* (3 Suppl):291S-9S.

Anker, M. (1996). Edible and Biodegradable Films and Coatings for Food Packaging–A Literature Review. *SIK- Report* No. 623.

Anon (1989). Official United States Standards for Oats. 7 CFR 810

Audrey H. Ensminger, 'Foods and Nutrition Encyclopedia', Volume 1, (Ensminger Pub Co, January 1983)ISBN 0-941218-05-8.

Ayres, J. C., Ocilvy, W. S., and Stewart, G. F. (1950). Post-mortem Changes in Stored meats, 1. Microorganisms Associated with Development of Slime on Eviscerated Cut-up Poultry. *Food Technology* 4, 199-205.

Bailey, H. and E. Bailey. 1976. Hortus Third. Cornell University MacMillan. N.Y. p 275.

Bansal, H. C., Shrivastava, K. N., Eggum, B. O. and Mehta, S. L. (1977). Nutritional Evaluation of High Protein Genotypes of Barley. *J Sci Food Agric* Feb; 28(2):157-60.

Bazzano, L. A., HE, J., Ogden, L. G., Loria, C. M., Whelton, P. K. (2003). Dietary Fiber Intake and Reduced Risk of Coronary Heart Disease in US Men and Women: The National Health and Nutrition Examination Survey I Epidemiologic Follow-up Study. *Arch Intern Med.* Sep 8;163(16):1897-904.

Behall, K. M., Scholfield, D. J., Hallfrisch, J. (2004). Diets Containing Barley Significantly Reduce Lipids in Mildly Hypercholesterolemic Men and Women. *Am J Clin Nutr.* Nov; 80 (5) : 1185-93.

Bird, H. R. (1943). The Nutritive Value of Eggs and Poultry Meat. Poultry and Egg National Board, *Nutrition Research Bull.* 5, Chicago, 111.

Bischoff-Ferrari, H. A., Dawson-Hughes. B., Willett, W. C. *et al.* (2004). Effect of Vitamin D on Falls: A Meta-Analysis. *JAMA*; 291:1999-2006.

Bishop, J. (1990). Oat Extract Called New Fat Fighter. *Wall St. J.* 85, No. 80, B-1 (4-20-90).

Booth, S. L., Broe, K. E., Gagnon, D. R. *et al.* (2003). Vitamin K Intake and Bone Mineral Density in Women and Men. *Am J Clin Nutr* 2003; 77:512-6.

Brady, D. E., Smith, F. H., Tucker, L. N., and Blumer, T. N. (1949) The Characteristics of Country Style Hams as Related to Sugar Content of Curing Mixture. *Food Res.* 14,303.

Brandenburg, A. H., Weller, C. L., and Testin, R. F. (1993). Edible Films and Coatings from Soy Protein. *J. Food* 4. Sci. 58:1086-1089.

Brand-Miller, J., Hayne, S., Petocz, P., Colagiuri, S. (2003). Low-glycemic Index Diets in the Management of Diabetes: A Meta-Analysis of Randomized Controlled trials. *Diabetes Care*, 26:2261-7.

Brehm, B. J., Seeley, R. J., Daniels, S.R., D'Alessio, D. A. (2003). A Randomized Trial Comparing a Very Low Carbohydrate Diet and a Calorie-Restricted Low Fat Diet on Body Weight and Cardiovascular Risk Factors in Healthy Women. *J Clin Endocrinol Metab*; 88:1617-23.

Broek, Van Den, C. J. H. (1965). Fish Canning. In: *Fish as Food* Vol. IV. Borgstrom, G. (ed.), Academic Press Inc. New York.

Brown, L., Rimm, E. B., Seddon, J. M. *et al.* (1999). A Prospective Study of Carotenoid Intake and Risk of Cataract Extraction in US Men. *Am J Clin Nutr*; 70:517-24.

Brown, L., Rosner, B., Willett, W. W., Sacks, F. M. (1999). Cholesterol-Lowering Effects of Dietary Fiber: A Meta-Analysis. *Am J Clin Nutr* 1999; 69:30-42.

Burgess, G. H. O. *et al.* (Eds) (1965) Fish Handling and Processing. HM Stationery office, Edinburgh. 390 pp.

Cappuccio, F. P., Elliott, P., Allender, P. S., Pryer, J., Follman, D. A., Cutler, J. A. (1995) Epidemiologic Association Between Dietary Calcium Intake and Blood Pressure: A Meta-Analysis of Published Data. *Am J Epidemiol* 1995; 142:935-45.

Casey, P. and Lorenz, K. (1977). Millet Functional and Nutritional Properties. Bakers Digest 51, No.1, 45-51

Chavan, J. T, and Kadam, S. S. (1989). Nutritional Improvement of Cereals by Fermentation. *Critical Rev. in Food Sci and Nutr.*

Cho, E., Seddon, J. M., Rosner, B., Willett, W. C. (2004). Hankinson SE. Prospective Study of Intake of Fruits, Vegetables, Vitamins, and Carotenoids and Risk of Age-Related Maculopathy. *Archives of Ophthalmology*; 122:883-92.

Coleby, R. (1959). The Effects of Irradiation on the Quality of Meat and Poultry. *International Journal of Applied Radiation and Isotopes* 6, 115-121.

Cramer, D. W., Harlow, B. L., Willett, W. C., *et al.* (1989). Galactose Consumption and Metabolism in Relation to the Risk of Ovarian Cancer. *Lancet*; 2:66-71.

Cramer, D. W. (1989). Lactase Persistence and Milk Consumption as Determinants of Ovarian Cancer Risk. *Am J Epidemiol*; 130:904-10.

Davis, L. L., and Coe, M. E. (1954). Bleeding of Chickens During Killing Operations. *Poultry Sci.* 33, 616-619.

Dawson, L. E., and Stadelman, W. J. (1960). Microorganisms and Their Control on Fresh Poultry Meat. NOM-7 Tech. Bull. 278, Mich. State Univ., E. Lans-ing, Mich.

De Lemos, M. L. (2001). Effects of Soy Phytoestrogens Genistein and Daidzein on Breast Cancer Growth. *Ann Pharmacother*; 35:1118-21.

Deane, D., and Commers, E. (1986). Oat Cleaning and Processing. In Oats: Chemistry and Technology. F. H. Webster, Ed. *Am. Assoc. Cereal Chemists*, St. Paul, MN.

Delaney, B., Nicolosi, R. J., Wilson, T. A. (2003). Beta-Glucan Fractions from Barley and Oats are Similarly Antiatherogenic in Hypercholesterolemic Syrian Golden Hamsters. *J Nutr*; Feb 133(2):468-75.

Desrosier, N.W. (1959). (Ed.) The Technology of Food Preservation. The AVI Publishing Co., Westport, Connecticut, USA.

Djousse L, Arnett DK, Coon H, Province MA, Moore LL, Ellison RC. (2004). Fruit and Vegetable Consumption and LDL Cholesterol: The National Heart, Lung, and Blood Institute Family Heart Study. *Am J Clin Nutr*; 79:213-7.

Draudt, H. W. (1963). The Meat Smoking Process, A Review. *Food Technol* 17, 85-90.

Dunker, C. F., Berman, M., Snider, G. G., and Tubiash, H. S. (1953). Quality and Nutritive Properties of Different Types of Commercially Cured Hams. III. Vitamin Content, Biological Value of Protein and Bacteriology. *Food Technol.* 7, 288.

Rhodehamel, E. J. (1992). "Overview of Biological, Chemical, and Physical Hazards," in M. D. Pierson and D. A. Corlett, Jr., eds., HACCP-Principles and Applications, A VI BookNan Nostrand Reinhold, New York, 1992, pp. 8-28.

Eggum, B. O. (1977). Nutritional Aspects of Cereal Proteins. In Genetic Diversity in Plants. A Muhammad, Ed. Plenum Press, NYC

Ensminger, A. H., Ensminger, M. E., Kondale, J. E. and Robson, J. R. K. Foods and Nutrition Encyclopedia. Pegus Press, Clovis, California.

Ensminger, A. H., Ensminger, M. K. J. *et. al.* (1986). Food for Health: A Nutrition Encyclopedia. Clovis, California: Pegus Press.

Erdman, J. W. (2000). Jr. AHA Science Advisory: Soy Protein and Cardiovascular disease: A Statement for Healthcare Professionals from the Nutrition Committee of the AHA. *Circulation*; 102:2555-9.

Fan, T. Y. and Tannenbaum, S. R. (1972). Stability of Nitroso Compounds. *J. Food Sci.* 37, 274.

Feskanich, D. Willett, W. C., Stampfer, M. J., and Colditz, G. A. (1996). Protein Consumption and Bone Fractures in Women. *Am J Epidemiol*; 143:472-9.

Feskanich, D., Willett, W. C., Stampfer, M. J., Colditz, G. A. (1997). Milk, Dietary Calcium, and Bone Fractures in Women: a 12-Year Prospective Study. *Am J Public Health*; 87:992-7.

Fortin and Francois, Editorial Director. The Visual Foods Encyclopedia. Macmillan, New York.

Foster, G. D., Wyatt, H. R., Hill, J.O. et al. (2003). A Randomized Trial of a Low-Carbohydrate Diet for Obesity. *N Engl J Med*; 348:2082-90.

Foster-Powell, K., Holt, S. H. (2002). Brand-Miller JC. International Table of Glycemic Index and Glycemic Load Values: 2002. *Am J Clin Nutr*; 76:5-56.

Fuchs, C. S., Giovannucci, E. L., Colditz, G. A. et al. (1999). Dietary Fiber and the Risk of Colorectal Cancer and Adenoma in Women. *N Engl J Med;* 340:169-76.

Rimm, E.B., Ascherio, A., Giovannucci, E., Spiegelman, D., Stampfer, M. J., Willett, W. C. (1996). Vegetable, Fruit, and Cereal Fiber Intake and Risk of Coronary Heart Disease Among Men. *JAMA*; 275:447-51.

Fuchs, C. S., Giovannucci, E. L., Colditz, G. A. *et al.* (1999). Dietary Fiber and the Risk of Colorectal Cancer and Adenoma in Women. *N Engl J Med*; 340:169-76.

Fung, T. T., Hu, F. B., Pereira, M. A. *et al.* (2002). Whole-Grain Intake and the Risk of Type 2 Diabetes: A Prospective Study in Men. *Am J Clin Nutr*; 76:535-40.

Gann, P. H., Ma, J., Giovannucci, E. *et al.* (1999). Lower Prostate Cancer Risk in Men with Elevated Plasma Lycopene Levels: Results of a Prospective Analysis. *Cancer Res*; 59:1225-30.

Ghorpade, V.M., U, H., Gennadios, A, and Hanna, M.A. (1995). Chemically Modified Soy Protein Film. Trans. of the ASAE 38:1805-1808.

Giddings, G. G., and Markikis, P. (1972). Characterization of the Red Pigments Produced from Ferrimyoglobin by Ionizing Radiation. *J. Food Sci.* 37, 361.

Gillies, M.T. (1971) Seafood Processing. Noyes Data Corporation, London.

Giovannucci, E., Ascherio, A., Rimm, E. B., Stampfer, M. J., Colditz, G. A., Willett, W.C. (1995). Intake of Carotenoids and Retinol in Relation to Risk of Prostate Cancer. *J Natl Cancer Inst*; 87:1767-76.

Giovannucci E, Goldin B. (1997). The Role of Fat, Fatty Acids, and Total Energy Intake in the Etiology of Human Colon Cancer. *Am J Clin Nutr*; 66:1564S-1571S.

Giovannucci, E., Rimm, E. B., Liu, Y., Stampfer, M. J., Willett, W. C. (2002). A Prospective Study of Tomato Products, Lycopene, and Prostate Cancer Risk. *J Natl Cancer Inst*; 94:391-8.

Giovannucci, E., Rimm, E. B., Wolk, A. *et al.* (1998). Calcium and Fructose Intake in Relation to Risk of Prostate Cancer. *Cancer Res*; 58:442-447.

Gray, J. I., Reddy, S. K, Price, J. F., Mandagere, A., and Wilkens, W. F. (1982). Inhibition of N-nitrosamines in Bacon. *Food Technol.* 36 (6), 39.

Gruttner, F. (1956). Handbook of Meat Processing, 5th Edition. Serger & Hempel. Braunschweig, Germany.

Guilbert, S. (1986). Technology and Application of Edible Protective Films. Food Packaging and Preservation: Theory and Practice. M. Mathlouthi, (Ed.), P 371-394. Elsevier, *Applied Science Publisher*, New York, NY.

Gutterson, M. (1972) Food Canning Techniques. Noyes Data Corporation, London.

Halton, T. L., Hu, F. B. (2004). The Effects of High Protein Diets on Thermogenesis, Satiety and Weight Loss: A Critical Review. *J Am Coll Nutr*; 23:373-85.

Health Claims: Soy Protein and Risk of Coronary Heart disease. *Code of Federal Regulations* 21CFR101.82 (2001).

Hersom, A.C. and Hulland, E.D. (1980). Canned Foods. *Churchill Livingstone*, London.

Hoagland, R. et al. (1947). Composition and Nutritive Value of Hams as Affected by Method of Curing. *Food Technol.* 1, 540.

Hooper M, Heighway-Bury R. (2001). Who Built the Pyramid? Cambridge, Mass.: Candlewick Press.

Hoseney, R. C., Andrews, D. J., and Clark, H. (1987). Sorghum and Pearl Millet. In Nutritional Quality of Cereal Grains. *American Society of Agronomy*, Madison, WI

Hu, F. B., Manson, J. E., Stampfer, M. J. (2001). Diet, Lifestyle, and the Risk of Type 2 Diabetes Mellitus in Women. *N Engl J Med*; 345:790-7.

Hu, F. B., Manson, J. E., Willett, W. C. (2001). Types of Dietary Fat and Risk of Coronary Heart Disease: A Critical Review. *J Am Coll Nutr*; 20:5-19.

Hu, F. B., Stampfer, M. J., Manson, J. E. (1999). Dietary Protein and Risk of Ischemic Heart Disease in Women. *Am J Clin Nutr*; 70:221-7.

Hulse, J. H., Laing, E. M., and Pearson, O. E. (1980). Sorghum and the Millets: Their Composition and Nutritive Value. *Academic Press*, NYC.

Hung, H. C., Joshipura, K. J., Jiang, R. (2004). Fruit and Vegetable Intake and Risk of Major Chronic Disease. *J Natl Cancer Inst*; 96:1577-84.

Hyman, J., Baron, J. A., Dain, B. J. (1998). Dietary and Supplemental Calcium and the Recurrence of Colorectal Adenomas. *Cancer Epidemiol Biomarkers Prev*; 7:291-5.

Jenkins, D. J., Kendall, C. W., Augustin, L, S. (2002). Glycemic Index: Overview of Implications in Health and Disease. *Am J Clin Nutr*; 76:266S-73S.

Jensen, M. K., Koh-Banerjee, P., Hu, F. B., Franz, M., Sampson, L., Gronbaek, M.and Rimm, E. B. (2004). Intakes of Whole Grains, Bran, and Germ and the Risk of Coronary Heart Disease in Men. *Am J Clin Nutr Dec*; 80(6):1492-9.

Johnsen, N. F., Hausner, H., Olsen, A., Tetens, I., Christenson, J., Knudsen, K. E., Overvad, K., Tjunneland, A. (2004). Intake of Whole Grains and Vegetables Determines the Plasma Enterolactone Concentration of Danish women. *J Nutr. Oct*;134(10):2691-7.

Jood, S. and Kalra, S. (2001). Chemical Composition and Nutritional Characteristics of Some Hull Less and Hulled Barley Cultivars Grown in India. *Nahrung*. Feb; 45 (1): 35-9.

John Eliot Coit (1915). Citrus Fruits: An Account of the Citrus Fruit Industry, with Special Reference to California Requirements and Practices and Similar Conditions. The Macmillan Company.

Kimball, Dan A. (June 30, 1999). Citrus Processing: A Complete Guide (2d ed.). New York: Springer. pp. 450. ISBN 0-8342-1258-7

Krebs, E. E., Ensrud, K. E., MacDonald, R. and Wilt, T. J. (2004). Phytoestrogens for Treatment of Menopausal Symptoms: A Systematic Review. *Obstet Gynecol*; 104:824-36.

Kreijkamp-Kaspers, S., Kok, L., Grobbee, D. E. Effect of Soy Protein Containing Isoflavones on Cognitive Function, Bone Mineral Density, and Plasma Lipids in Postmenopausal Women: A Randomized Controlled Trial. *JAMA*; 292:65-74.

Krinsky, N. I,, Landrum, J. T., Bone, R. A. (2003). Biologic Mechanisms of the Protective Role of Lutein and Zeaxanthin in the Eye. *Annu Rev Nutr*; 23:171-201.

Kris-Etherton, P. M., Harris, W. S., Appel, L. J. (2002). Fish Consumption, Fish Oil, Omega-3 Fatty Acids, and Cardiovascular Disease. *Circulation*; 106:2747-57.

Kritz-Silverstein, D., Von Muhlen, D. Barrett-Connor, E., Bressel, M. A. (2003). Isoflavones and Cognitive Function in Older Women: The Soy and Postmenopausal Health in Aging (SOPHIA) Study. *Menopause*; 10:196-202.

Kronenberg, F., Fugh-Berman, A. (2002). Complementary and Alternative Medicine for Menopausal Symptoms: A Review of Randomized, Controlled Trials. *Ann Intern Med* 2002; 137:805-13.

Kushi L, Giovannucci E. (2002). Dietary Fat and Cancer. *Am J Med*; 113 Suppl 9B:63S-70S.

Leaf, A., Kang, J. X., Xiao, Y. F., Billman, G. E. (2003). Clinical Prevention of Sudden Cardiac Death by n-3 Polyunsaturated Fatty Acids and Mechanism of Prevention of Arrhythmias by n-3 Fish Oils. *Circulation* 2003; 107:2646-52.

Lembo, A., Camilleri, M. (2003). Chronic Constipation. *N Engl J Med* 2003; 349:1360-8.

Liese, A. D., Roach, A. K., Sparks, K. C., Marquart, L., D'Agostino, R. B., Jr., Mayer-Davis, E. J. (2003). Whole-Grain Intake and Insulin Sensitivity: The Insulin Resistance Atherosclerosis Study. *Am J Clin Nutr*; 78:965-71.

Liu, S., Willett, W. C., Stampfer, M. J. (2000). A Prospective Study of Dietary Glycemic Load, Carbohydrate Intake, and Risk of Coronary Heart Disease in US Women. *Am J Clin Nutr*; 71:1455-61.

Liu, S., Willett, W. C. (2002). Dietary Glycemic Load and Atherothrombotic Risk. *Curr Atheroscler Rep*; 4:454-61.

Liu, R. H. (2004). New Finding may be Key to Ending Confusion Over Link Between Fiber, Colon Cancer. *American Institute for Cancer Research Press Release*, November 3, 2004.

Liu, S. (1998). Insulin Resistance, Hyperglycemia and Risk of Major Chronic Diseases—A Dietary Perspective. *Proc Nutrit Soc Austral*; 22:140.

Lockhart, H. B., and Hurt, H. D. (1986). Nutrition of Oats. In Oats: Chemistry and Technology. F. H. Webster, Ed. *Am. Assoc. Cereal Chemists*, St. Paul, MN.

Lockhart, H. B. and Nesheim, R. O. (1978). Nutritional Quality of Cereal Grains. In Cereals 78: Better Nutrition for the World's Millions. Y. Pomeranz, Ed. *Am. Assoc. Cereal Chemists*, St. Paul, MN.

Macarthur-Grant, L. A. (1986). Sugars and Non-Starchy Polysaccharides in Oats. In Oats: Chemistry and Technology. F. Webster, Ed. *Am. Assoc. Cereal Chem.* St.

Mackenzie, D. S. (1966). Prepared Meat Product Manufacturing, 2nd Edition. *American Meat Institute,* Chicago, II.

Manson, J. E., Hsia, J., Johnson, K. C. *et al*. (2003). Estrogen Plus Progestin and the Risk of Coronary Heart Disease. *N Engl J Med*; 349:523-34.

Martinez, M.E., Willett, W. C. (1998). Calcium, Vitamin D, and Colorectal Cancer: A Review of the Epidemiologic Evidence. *Cancer Epidemiol Biomarkers Prev*; 7:163-8.

Matz, S. A. (1959). The Chemistry and Technology of Cereals as Food and Feed. AVI Publishing Co., Westport, CT (Ed).

Matz, S. A. (1969). Cereal Science. AVI Publishing Co., Westport, CT.

McCullough, M. L., Feskanich, D., Stampfer, M. J. et al. (2002). Diet Quality and Major Chronic Disease Risk in Men and Women: Moving Toward Improved Dietary Guidance. *Am J Clin Nutr*; 76:1261-71.

McKeown, N. M., Meigs, J. B., Liu, S., Saltzman, E., Wilson, P. W., Jacques, P. F. (2004). Carbohydrate Nutrition, Insulin Resistance, and the Prevalence of the Metabolic Syndrome in the Framingham Offspring Cohort. *Diabetes Care*; 27:538-46.

McKeown NM, Meigs JB, Liu S, Wilson PW, Jacques PF. Whole-Grain Intake is Favorably Associated with Metabolic Risk Factors for Type 2 Diabetes and Cardiovascular Disease in the Framingham Offspring Study. *Am J Clin Nutr*; 76:390-8.

McMichael-Phillips, D. F., Harding, C., Morton, M. *et al.* (1998). Effects of Soy-Protein Supplementation on Epithelial Proliferation in the Histologically Normal Human Breast. *Am J Clin Nutr*; 68:1431S-1435S.

Messina, M., Gardner, C. and Barnes, S. (2002). Gaining Insight into the Health Effects of Soy but a Long Way Still to go: Commentary on the Fourth International Symposium on the Role of Soy in Preventing and Treating Chronic Disease. *J Nutr*; 132:547S-551S.

Miller, B. S., and Joghson, J. A. (1954). A Review of Methods for Determining the Quality of Wheat and Flour for Breadmaking. Kans. Agr. Expt. Sta. Tech. Bull. 76.

Moeller, S. M., Taylor A, Tucker, K. L. *et al*. (2004). Overall Adherence to the Dietary Guidelines for Americans is Associated with Reduced Prevalence of Early Age-Related Nuclear Lens Opacities in Women. *J Nutr*; 134:1812-9.

Morrison, W. R. (1988). Lipids. In Wheat: Chemistry and Technology, Vol. I, Third Edition. Y. Pomeranz, Ed. *Am. Assoc. Cereal Chemists*, St. Paul, MN, Mushroom Cultivator: A Practical Guide to Growing Mushrooms at Home (1983) ISBN 0961079800; National Advisory Committee on Microbiological Criteria for Foods, "Hazard Analysis and Critical Control Point Principles and Application Guidelines," *J. Food Prot.* 61,762-775(1998).

Morton, J., Fruits of Warm Climates (1987) Miami, FL, pp. 134–142.

Neufeld, C. H. H., Weinstein, N. E., and Mecham, D. K. (1957). Studies on the Preparation and Keeping Quality of Bulgur. *Cereal Chern.* 34, 360-370.

Nicolosi, E.; Deng, Z. N.; Gentile, A.; La Malfa, S.; Continella, G.; Tribulato, E. (2000). "Citrus Phylogeny and Genetic Origin of Important Species as Investigated by Molecular Markers". TAG *Theoretical and Applied Genetics* 100 (8): 1155–1166.

Norbaek, R., Brandt, K. and Kondo, T. (2000). Identification of Flavone C-glycosides Including a New Flavonoid Chromophore from Barley Leaves (Hordeum vulgare L.) by Improved NMR Techniques. *J Agric Food Chem*. May; 48 (5) : 1703-7.

Orth, R. A. and Shellenberger, J. A. (1988). Origin, Production, and Utilization of Wheat. In Wheat: Chemistry and Technology, Vol. I, Third Ed. Y. Pomeranz. *Am. Ass De. Cereal Chemists*, St. Paul, MN.

Owusu, W., Willett, W. C., Feskanich, D., Ascherio, A., Spiegelman, D. and Colditz, G. A. (1997). Calcium Intake and the Incidence of Forearm and Hip Fractures Among Men. *J Nutr*; 127:1782

Papadimitropoulos, E., Wells, G., Shea, B. *et al.* (2002). Meta-Analyses of Therapies for Postmenopausal Osteoporosis. VIII: Meta-Analysis of the Efficacy of Vitamin D Treatment in Preventing Osteoporosis in Postmenopausal Women. *Endocr Rev*; 23:560-9.

Paton, D. (1986). Oat Starch: Physical, Chemical, and Structural Properties. In Oats: Chemistry and Technology. F. H. Webster, Ed. *Am. Assoc. Cereal Chemists*, St. Paul, MN

Paul, M. N. and Matz, S. A. (1972). Cereal Science. A VI Publishing Co., Westport, CT

Pereira, M. A., Liu, S. (2003). Types of Carbohydrates and Risk of Cardiovascular Disease. *J Womens Health (Larchmt);* 12:115-22.

Pereira, M. A., O'Reilly, E., Augustsson, K. *et al.* (2004). Dietary Fiber and Risk of Coronary Heart Disease: A Pooled Analysis of Cohort Studies. *Arch Intern Med* 2004; 164:370-6.

Peterson, D. M., Senturia, J., Youngs, V. L., and Schrader, L. E. (1975). Elemental Composition of Oat Groats. *J. Agr. Food Chemistry* 23,9-13.

Piggot-G.M., Who is the 21st Century Consumer? *Infofish International*, 1/94.

Reaven, G. M. (2003). Insulin Resistance/Compensatory Hyperinsulinemia, Essential Hypertension, and Cardiovascular Disease. *J Clin Endocrinol Metab*; 88:2399-403.

Rimm, E. B., Ascherio, A., Giovannucci, E., Spiegelman, D., Stampfer, M. J., Willett, W. C. (1996). Vegetable, Fruit, and Cereal Fiber Intake and Risk of Coronary Heart Disease Among Men. *JAMA*; 275:447-51.

Samaha, F. F., Iqbal, N., Seshadri, P., A Low-Carbohydrate as Compared with a Low-Fat Diet in Severe Obesity. *N Engl J Med*; 348:2074-81.

Schrickel, D. J. (1986). Oats Production, Value, and Use. In Oats: Chemistry and Technology. F. H. Webster, Ed. *Am. Assoc. Cereal Chemists*, St. Paul, MN.

Schulze, M.. B., Liu, S., Rimm, E. B., Manson, J. E., Willett, W. C., Hu, F. B. (2004). Glycemic Index, Glycemic Load, and Dietary Fiber Intake and Incidence of Type 2 Diabetes in Younger and Middle-Aged Women. *Am J Clin Nutr*; 80:348-56.

Shellenberger, J. A. (1969). Wheat. In Cereal Science. S. A Matz, Ed. AVI Publishing Co., Westport, CT

Shellenberger, J. A. (1980). Advances in Milling Technology. In Advances in Cereal Science and Technology 3. Y. Pomeranz, Ed. *Am. Assoc. Cereal Chemists*, St. Paul, MN

Shukla, T. P. (1975). Chemistry of Oats: Protein Foods and Other Industrial Products. Crit. *Rev. Food Sci. Nutr.* 6, 383.431

Sieri S, Krogh V, Pala V, *et al.* (2004). Dietary Patterns and Risk of Breast Cancer in the ORDET Cohort. *Cancer Epidemiol Biomarkers Prev*; 13:567-72.

Smith-Warner SA, Spiegelman D, Adami HO, et al. Types of Dietary Fat and Breast Cancer: A Pooled Analysis of Cohort Studies. *Int J Cancer*; 92:767-74.

Stanton, T. R. (1953). Production, Harvesting, Processing, Utilization, and, Economic Importance of Oats. Econ. Bot. 7, 43-64.

Stern, L., Iqbal, N., Seshadri, P. et al. (2004). The Effects of Low-Carbohydrate Versus Conventional Weight Loss Diets in Severely Obese Adults: One-year Follow-up of a Randomized Trial. *Ann Intern Med*; 140:778-785.

Tsai, C. J., Leitzmann, M. F., Willett, W. C., Giovannucci, E. L. (2004). Long-Term Intake of Dietary Fiber and Decreased Risk of Cholecystectomy in Women. *Am J Gastroenterol. Jul*; 99(7):1364-70.

USDA National Nutrient Database for Standard Reference, Release 17. *U.S. Department of Agriculture, Agricultural Research Service*: 2004.

Vainio H, Bianchini F. (2003). IARC Handbooks of Cancer Prevention: Fruit and Vegetables. Vol. 8. Lyon, France.

Van Horn, L. (1997). Fiber, Lipids, and Coronary Heart Disease. A Statement for Healthcare Professionals from the Nutrition Committee, American Heart Association. *Circulation*; 95:2701-4.

Webber, Herbert John; rev Walter Reuther and Harry W. Lawton; Willard Hodgson (1967–1989) [1903]. Citrus Sinensis – *Encyclopedia of Life*. EOL. Retrieved on 2011-10-02.

Webber, Herbert John; rev Walter Reuther and Harry W. Lawton (1967–1989). The Citrus Industry. Riverside CA: University of California Division of Agricultural Sciences. Home Fruit Production – Oranges, **Julian W. Sauls**, Ph.D., Professor & Extension Horticulturist, Texas Cooperative Extension (December, 1998), aggie-Horticulture.tamu.edu

Weber P. (2001). Vitamin K and Bone Health. *Nutrition*; 17:880-7.

White, L. R., Petrovitch, H., Ross, G. W. *et al.* (2000). Brain Aging and Midlife Tofu Consumption. *J Am Coll Nutr*; 19:242-55.

Whittle-K.J.. Hardy-R. and Hobbs-G. (1990). Chilled Fish and Fishery Products, In: Chilled Foods, Gormley- TR (ed), Commission of the European Communities, *Elsevier Applied Science*. London.

Willett, W., Manson, J. and Liu, S. (2002). Glycemic Index, Glycemic Load, and Risk of Type 2 Diabetes. *Am J Clin Nutr*; 76:274S-80S.

Willett, W. C., MacMahon, B. (1984). Diet and Cancer—An Overview. *N Engl J Med*; 310:633-8.

Willett, W. C., MacMahon, B. (1984). Diet and Cancer—an Overview (Second of Two Parts). *N Engl J Med*; 310:697-703.

Willett, W. C., Stampfer, M. J., Manson, J. E. et al. (1993). Intake of Trans Fatty Acids and Risk of Coronary Heart Disease Among Women. *Lancet*; 341:581-5.

Wood, R. (1988). The Whole Foods Encyclopedia. New York, NY: Prentice-Hall Press;.

Yancy, W. S., Olsen, M. K., Guyton, J. R., Bakst, R. P., Westman, E. C. (2004). A Low-Carbohydrate, Ketogenic Diet Versus a Low-Fat Diet to Treat Obesity and Hyperlipidemia: A Randomized, Controlled Trial. *Ann Intern Med*; 140:769-777.

Web References

http://en.wikipedia.org

http://en.wikipedia.org/wiki/Cantharellus

http://en.wikipedia.org/wiki/Mushroom

http://lib.ucr.edu/agnic/webber/

http://lib.ucr.edu/agnic/webber/Vol1/Chapter4.html.

http://mushrooms.simons-rock.edu/

http://plants.montara.com/mushrooms/MListPages/MFamPages/cantharella.html

http://plants.usda.gov/java/profile?symbol=CISI3.

http://vm.cfsan.fda.gov/~lrd/haccp.html

http://vm.cfsan.fda.gov/~lrd/haccp.html

http://vm.cfsan.fda.gov/list.html

http://vm.cfsan.fda.gov/list.html

http://warp.nal.usda.gov/fnic/HEI/hlthyeat.pdf

http://warp.nal.usda.gov/fnic/HEI/hlthyeat.pdf

http://web.archive.org/web/20110221180126/

http://wm.cfsan.fda.gov/~dms/haccp-2.html

http://wm.cfsan.fda.gov/~dms/haccp-2.html

http://www.arhomeandgarden.org/plantoftheweek/articles/hardy_orange_2-9-07.htm.

http://www.ars-grin.gov

http://www.ars-grin.gov/cgi-bin/npgs/html/taxon.pl?10782

http://www.bluewillowpages.com/mushroomexpert/

http://www.bright.net/~wildwood

http://www.cfaitc.org/factsheets/pdf/CitrusFruits.pdf

http://www.citrusvariety.ucr.edu/citrus/sweet_oranges.html

http://www.conservation.state.mo.us/nathis/mushrooms/chanterelles/

http://www.consumersunion.org/food/0908fooddc998.htm.

http://www.darzau.de/en/projects/hulless_food_barley.htm

http://www.deliciousorganics.com/recipes/mushrooms.htm

http://www.ent.iastate.edu/imagegal/plantpath/barley/

http://www.erowid.org/plants/mushrooms/mushrooms_info12.shtml

http://www.etymonline.com/index.php?term=orange.

http://www.grains.org/grains/barley.html

http://www.hsph.harvard.edu/nutritionsource/pyramids.html

http://www.mushroomcouncil.org/

http://www.mushworld.com/

http://www.mykoweb.com/

http://www.mykoweb.com/CAF/species/Cantharellus_cibarius.html

http://www.mykoweb.com/CAF/species/Cantharellus_subalbidus.html

http://www.mykoweb.com/CAF/species/Cantharellus_tubaeformis.html

http://www.mykoweb.com/CAF/species/Craterellus_cornucopioides.html

http://www.mykoweb.com/CAF/species_index.html#C

http://www.mykoweb.com/cookbook/chanterelle.html

http://www.nal.usda.gov/fnic/foodborne/foodborn.htm

http://www.nal.usda.gov/fnic/foodborne/foodborn.htm

http://www.namyco.org/

http://www.ncbi.nlm.nih.gov/Taxonomy/Browser/wwwtax.cgi?mode=Undef&id=36065 &lvl=3&keep=1&srchmode=1&unlock

http://www.plantphysiol.org/cgi/reprint/20/1/3.pdf.

http://www.shroomery.org/

http://www.state.ak.us/dec/deh/seafood/haccpweb.htm

http://www.state.ak.us/dec/deh/seafood/haccpweb.htm

http://www.ubcbotanicalgarden.org/potd/2007/02/citrus_sinensis_cara_cara.php.

http://www.washingtonbarley.org/recipes.html

http://www.wild-harvest.com

http://www-seafood.ucdavis.edu/

http://www-seafood.ucdavis.edu/

kin-jay.blogspot.com

http://www.fruitspulp.com/images/guava_juice.jpg

http://www.recipetips.com/images/glossary/n/nectar_mango.jpg

http://www.21food.com/userImages/pastoralchina/pastoralchina$11614216.jpg

http://www.indiamart.com/paramount/pcat-gifs/products-small/guava-pulp.jpg